The Anthropology of Disasters in Latin America

This book offers anthropological insights into disasters in Latin America. It fills a gap in the literature by bringing together national and regional perspectives in the study of disasters.

The book essentially explores the emergence and development of anthropological studies of disasters. It adopts a methodological approach based on ethnography, participant observation, and field research to assess the social and historical constructions of disasters and how these are perceived by people of a certain region. This regional perspective helps assess long-term dynamics, regional capacities, and regional-global interactions on disaster sites. With chapters written by prominent Latin American anthropologists, this book also considers the role of the state and other nongovernmental organizations in managing disasters and the specific conditions of each country, relative to a greater or lesser incidence of disastrous events.

Globalizing the existing literature on disasters with a focus on Latin America, this book offers multidisciplinary insights that will be of interest to academics and students of geography, anthropology, sociology, and political science.

Virginia García-Acosta is a Mexican social anthropologist and historian, as well as a teacher and researcher since 1973 in CIESAS, Mexico. Her research relates to disaster and risk from a historical-anthropological perspective, focused in Mexico and Latin America. Her most recent book was *Les Catastrophes et l'interdisciplinarité* (Louvain, 2017), and *History and Memory of Hurricanes and Other Hydrometeorological episodes in Mexico: Five Centuries* (Mexico) is forthcoming.

Routledge Studies in Hazards, Disaster Risk and Climate Change

Series Editor: Ilan Kelman
Professor of Disasters and Health at the Institute for Risk and Disaster Reduction (IRDR) and the Institute for Global Health (IGH), University College London (UCL)

This series provides a forum for original and vibrant research. It offers contributions from each of these communities as well as innovative titles that examine the links between hazards, disasters and climate change, to bring these schools of thought closer together. This series promotes interdisciplinary scholarly work that is empirically and theoretically informed, with titles reflecting the wealth of research being undertaken in these diverse and exciting fields.

For more information about this series, please visit: www.routledge.com/series/HDC

The Anthropology of Disasters in Latin America

State of the Art

Edited by Virginia García-Acosta

Routledge
Taylor & Francis Group

LONDON AND NEW YORK

First published 2020
by Routledge
2 Park Square, Milton Park, Abingdon, Oxon OX14 4RN

and by Routledge
52 Vanderbilt Avenue, New York, NY 10017

Routledge is an imprint of the Taylor & Francis Group, an informa business

First issued in paperback 2021

British Library Cataloguing-in-Publication Data
A catalogue record for this book is available from the British Library

Library of Congress Cataloging-in-Publication Data
A catalog record for this book has been requested

ISBN: 978-1-138-58145-6 (hbk)
ISBN: 978-1-03-208199-1 (pbk)
ISBN: 978-0-429-50672-7 (ebk)

Typeset in Times New Roman
by Apex CoVantage, LLC

Contents

Maps

Contributors

Altez, Rogelio. Universidad Central de Venezuela (Central University of Venezuela).
Anthropologist and Historian. Titular Professor (School of Anthropology, Central University of Venezuela). Doctor in History (University of Seville). Awards: National Prize of History (National Academy of History, Venezuela, 2011); Nuestra América Prize (CSIC-University of Seville, 2015); Special PhD Award (University of Seville, 2017). Research stays and Visiting Professor in Spain, Mexico, France, Chile, and Colombia. More recent book: *Historia de la Vulnerabilidad en Venezuela. Siglos XVI-XIX* (Madrid, CSIC-Universidad de Sevilla, 2016). He has created the subject *Anthropology of Disasters* in 2009 (School of Anthropology, Central University of Venezuela) the first on the topic in Latin America.

Barrios, Roberto E. Southern Illinois University Carbondale, USA.
He is a Guatemalan-born disaster anthropologist who has conducted ethnographic research in Central America, Mexico, and the United States during the last 20 years. His research has focused on the modernist and neoliberal assumptions of post-disaster community reconstruction programs and the ways disaster survivors navigate, interpret, or challenge these recovery processes. He has published in the *Annual Review of Anthropology, Human Organization, Anthropology News, Identities: Global Studies in Culture and Power* and a number of edited volumes. He is the author of *Governing Affect: Neoliberalism and Disaster Reconstruction* (University of Nebraska Press, 2017). He has served as the lead co-chair of the Risk and Disaster Topical Interest Group of the Society for Applied Anthropology.

Batres, Carlos. Southern Illinois University Carbondale, USA.
Originally from Guatemala City, he is currently a PhD candidate in Cultural Anthropology at Southern Illinois University at Carbondale. He is also a trained archaeologist, whose research has focused on environmental conditions and technological developments that enabled the earliest settlements in the American continent and the role of these factors in population displacement. His present research interest as a sociocultural anthropologist is in artificial intelligence (AI) and robotics, specifically in the definitions emerging from the relationship between people and machines, and the impact of this relationship on human mobility related to the emergence of specialized technological communities.

Bravo Alarcón, Fernando. Pontificia Universidad Católica del Perú (Pontifical Catholic University of Peru).
Peruvian sociologist, Fernando is a Magister in Environmental Development and also in Political Science and has finished his doctoral studies in Anthropology in the Pontifical Catholic University of Peru, where he is Professor. His areas of interest include climate change, environmental issues, and disasters. He has published various academic articles on these topics, among them are these: "Environmental issues in the anthropological theory" in the *Revista Peruana de Antropología*, "Social research on climate change" in the magazine *Argumentos*, "Complex aspects of the environmental awareness in Peru" in the journal *Socialismo y Participación*. He is the author of the book *The Faustian Pact of La Oroya: The Right to the Beneficial Pollution* (INTE-PUCP, 2015).

Camargo, Alejandro. Universidad del Norte in Barranquilla, Colombia (North University, Colombia).
Colombian anthropologist and geographer. He holds a PhD in Geography from Syracuse University and was a postdoctoral researcher at the department of Geography at the Université de Montréal. He is currently Assistant Professor in the Department of History and Social Sciences at the Universidad del Norte in Barranquilla, Colombia. He is broadly interested in water-related disasters, agrarian political economy, and the political ecology of rivers and wetlands. Forthcoming: Camargo, A. and Cortesi, L., "Flooding water and society", *Wiley Interdisciplinary Reviews: Water*. In preparation: "Natural disasters and risk", B. Bustos, D. Ojeda, G. García López, F. Milanez and S.E Di-Mauro, eds., *Handbook of Latin America and the Environment* (Routledge).

Faas, A.J. San José State University, USA.
PhD in Anthropology, University of South Florida. He is Associate Professor of Anthropology at San José State University. He studies disasters, environmental crises, and displacement and resettlement, principally in Mexico, Ecuador, and the United States. His work focuses on social organization and economy at multiple levels of scale, postcolonial statecraft and practice, interventions of nongovernmental organizations, and the (re)production of subjectivities and memories. His work has appeared in *Human Organization, Annals of Anthropological Practice, Disasters, Human Nature, International Journal of Disaster Risk Reduction, Disaster Prevention and Management, Economic Anthropology, Journal of Latin American and Caribbean Anthropology, Ethnology, Development In Practice*, and several edited volumes.

García-Acosta, Virginia. Centro de Investigaciones y Estudios Superiores en Antropología Social, Mexico. (Center for Research and Advanced Studies in Social Anthropology, CIESAS).
Mexican Social Anthropologist and Historian. Professor and researcher at CIESAS since 1973. Her research relates to disaster and risk from a historical-anthropological perspective. Some of her recent publications: "Building on the past. Disaster Risk Reduction including Climate Change Adaptation in the *Longue Durée*" (Routledge, 2017), "Unnatural disasters and the Anthropocene: lessons learnt from

anthropological and historical perspectives in Latin America" (Il Sileno Edizioni, 2019), and the book with A. Musset: *Les Catastrophes et l'interdisciplinarité: dialogues, regards croisés* (Academia-L'Harmattan, 2017). Forthcoming is the book *Historia y memoria de los huracanes y otros episodios hidrometeorológicos en México: Cinco siglos* (*History and Memory of Hurricanes and Other Hydrometeorological Episodes in Mexico: Five Centuries*), with R. Padilla.

Kelman, Ilan. University College London, England and University of Agder, Kristiansand, Norway.
He is Professor of Disasters and Health at University College London, England, and a Professor II at the University of Agder, Kristiansand, Norway. His overall research interest is linking disasters and health, including the integration of climate change into disaster research and health research. That covers three main areas: a) disaster diplomacy and health diplomacy www.disasterdiplomacy.org; b) island sustainability involving safe and healthy communities in isolated locations www.islandvulnerability.org; and c) risk education for health and disasters www.riskred.org. www.ilankelman.org and Twitter/Instagram *@IlanKelman*.

Murgida, Ana María. Universidad de Buenos Aires, Argentina (University of Buenos Aires).
Argentinian Sociocultural Anthropologist, lecturer and researcher at the University, and senior consultant in national and international level. Her experience as a leader on interdisciplinary project teams is centered on social risk studies in contexts of global change, risk management, and on science-policy interface, on different ecosystems and with social actors as aboriginal and creole communities, governmental institutions, etc. The results of her work have been reflected in academic and institutional papers.

Oliver-Smith, Anthony. Professor Emeritus of Anthropology at the University of Florida, Gainesville.
He held the Greenleaf Chair of Latin American Studies at the Stone Center for Latin American Studies at Tulane University (2008) and the Munich Re Foundation Chair on Social Vulnerability at the United Nations University Institute on Environment and Human Security in Bonn, Germany (2005–9). He was awarded the *Bronislaw Malinowski Award of the Society for Applied Anthropology in 2013* for lifetime achievement for his work in disaster studies and resettlement research. He has done anthropological research and consultation on issues relating to disasters and involuntary resettlement in Peru, Honduras, India, Brazil, Jamaica, Mexico, Panama, Colombia, Japan, and the United States.

Radovich, Juan Carlos. Universidad de Buenos Aires, Argentina (University of Buenos Aires).
He was born and raised in Buenos Aires City, Argentina. PhD in Social Anthropology, University of Buenos Aires (UBA), and he is Professor at its Anthropology Department. Senior Research Fellow at the National Council for Scientific

and Technological Research (CONICET). He is codirector of the Program "Ethnicities and Territories in redefinition" at the Institute of Anthropological Sciences (ICA) at the UBA. His research skills are as follows: Rural Anthropology, Applied Anthropology, Social Impact of Tourism among Indigenous Populations, Big Dams' Social Impacts, Population Resettlements, Social Effects of Oil and Gas Production, Rural-Urban Migration, Indigenous Policies, Indigenous Political Movements in Argentina, Ethnic Conflicts in Modern World, Racism and Discrimination in Contemporary Society.

Taddei, Renzo. Universidade Federal de São Paulo, Brazil (Federal University of Sao Paulo).
Professor of anthropology at the Federal University of Sao Paulo, Brazil. Dr. Taddei has earned his doctoral degree in anthropology from Columbia University. He has served as Visiting Professor at Yale, Duke, and the University of the Republic of Uruguay. Dr. Taddei's work focuses on environmental conflicts and traditional environmental knowledge in South America.

Taks, Javier. Universidad de la República, Uruguay (University of the Republic of Uruguay).
Uruguayan anthropologist, senior lecturer at the Universidad de la República, Uruguay. Dr. Taks has earned his doctoral degree in social anthropology from The University of Manchester, UK. Since 2013 he became chairman of the UNESCO Chair on Water and Culture. His work focuses on water governance, energy systems, development, and environmental conflicts in South America. He has made consultancy on land planning for watershed management.

Foreword

To try to understand disasters means to try to understand people. What are the beliefs, belief systems, values, interests, fundamental drivers, and fundamental inhibitors which lead to our actions, choices, and behaviors all combining to reduce, create, or ignore disaster risk? What are the meanings and expressions of cultures, customs, ethics, philosophies, norms, and outliers, some of which tackle vulnerability, others which increase it, and some of which have no effect?

What motivates some of us to dive deeply into the meanings, connotations, and interpretations of basic concepts such as "disaster", "risk", and "vulnerability", while others accelerate along the endless spiral of pointlessly complex and increasingly meaningless nomenclature such as "resilient adaptive capacity", "transformational change", and "social-ecological systems"? While some seem to be content with exploring empathy – identifying with others, such as when they suffer – others push far beyond, towards caring about improving people's situations. How do we explain the reasons for many still assuming that climate change is a principal cause of disasters, migration, and conflict, compared to those who dissect the root causes of politics and power?

The answers to all these questions can be found by understanding people, individually and collectively. Not exclusively, though. We still need to probe seismic statistics, improve our hydrodynamic modeling of storm surge, do better at field surveys of structural failure under volcanic blast and ash loading, and improve our laboratory and on-site empirics of attempts at malaria vaccines.

So much still comes back to people. Which means the importance of studying people. This volume strides forward by focusing on anthropology.

Studying people means celebrating our diversity. We are too well aware of how much research is dominated by English writing and Anglophone thinking. Yet we must be cautious since every culture and language has positive, negative, and neutral aspects. The key is combining the positive points from everyone in order to diminish and overcome the negative baggage which we all bring. This volume proffers exactly this by highlighting Latin America – and especially through many authors writing about their own countries and their own contexts. Not that Latin America is a single homogenous entity! Being geographically huge and rich in its own diversity means plenty of positive, negative, and neutral aspects about which to learn.

We can and should learn from the authors here, sifting through their advice for our own locations, inviting them to help us study ourselves, and then offering in return our own teaching, recommendations, knowledge, and wisdom. We create an exchange of mutual teaching and learning – of learning and teaching about people and disasters. The mammoth task of compiling such a volume and nursing it to completion, with deepest appreciation to the editor, is in itself a process of starting this exchange in order to learn and teach about ourselves.

The *Routledge Studies in Hazards, Disaster Risk and Climate Change Series* welcomes this book as a much-needed contribution about people and disasters from diverse perspectives within Latin America and anthropology. It forges a beginning point for discussion, showing us how little each of us knows and how much we can learn from so many people around us. It is not about merely trying to understand people in order to try to understand disasters. It is about succeeding so that we help people deal with and ultimately avoid disasters, making life and the world safer for us all.

Ilan Kelman

Prologue

I write this prologue as an anthropologist who has enjoyed the privilege of having observed and experienced the evolution of an intellectual tradition that was and is both pioneering and productive in the field of disaster research in Latin America and the world. In effect, Latin America has provided the context for theoretically imaginative, methodologically diverse anthropological studies of hazards, risks and disasters, all the while maintaining relevance to both policy and practice in the region and, by extension, the world. This book, marking an important benchmark in the progress of that research field, is thus an expression of the continuing vitality of that tradition.

The paradigm shift that began in the mid-1970s, reorienting the focus of research and policy eventually from hazards, impacts, and reconstruction toward the social construction of risk and disasters, began with researchers from and working in the developing world. Out of that paradigm shift, the seeds of a Latin American tradition in social scientific study of risk and disasters emerged. Thus, when I began my own research in mid-1970 into a disaster that occurred in the field site I had been preparing to study in the north-central Andes of Peru, I thought my first steps prior to entering the field would naturally be a social science literature search on disasters in Peru. However, finding none, I broadened my scope to Latin America, which proved equally fruitless. The pioneering work of Fernando Ortiz, a Cuban anthropologist, unfortunately was not in the card catalogues of North American libraries in 1970. Finally, in desperation, I expanded my search to encompass the developing world in general. There I found roughly a dozen articles by anthropologists who happened to be doing fieldwork on some other topic when natural hazards such as volcanic eruptions, typhoons, or hurricanes had occurred during their field work in the South Pacific region. One other anthropological contribution that must be mentioned was Anthony F. C. Wallace's study of the Worcester, Massachusetts, tornado in 1953, in which he laid the groundwork for the temporal and spatial analysis of the social dimensions of disaster impact. Although Fernando Ortiz, the true pioneer of the anthropology of disasters, published *El huracán: su mitología y sus símbolos* in 1947, his work was not broadly recognized until well into the 1980s. It is thus more than a little ironic that the true pioneer in the study of disasters in anthropology is a Latin American, a Cuban, who went relatively unrecognized until the 1980s, probably more for political and cultural reasons than academic or scientific.

Consequently, for almost two decades I felt quite lonely as an anthropologist studying disasters in Latin America. As a matter of fact, I knew of only one other anthropologist, William Torry, who was interested in the field, but he was working largely in Africa on issues of famine, but writing extensively about other disasters in the Middle East and India. In general it was not until the 1990s that the anthropological study of disasters began to emerge as a recognized anthropological research focus. Hence, when I was invited in 1991 to attend a conference called Disaster: Vulnerability and Development at the Royal Geographical Society (RGS) in London, I still went with little expectation that there would be much on Latin America, and even less on anthropology. It was there in the august and imposing lecture hall of the RGS, encircled by a balcony level, dark wooden band bearing the foot high names of famous scientists and explorers, that I met fellow presenters Andrew Maskrey and Allan Lavell. When each of us, completely unknown to the other, found out that we were each working on disasters in Latin America, it was a little like discovering long-lost brothers who had been separated at birth. Allan, a geographer, and Andrew, an urban planner, and I, two Englishmen and a North American from the US, but all three with deep and longstanding ties to Latin America, were all working on exploring issues of risk, vulnerability and the root causes of disaster. When the conference was over, we enthusiastically agreed to stay in contact.

Not six months later at the University of California, Los Angeles, the UCLA International Conference on the Impact of Natural Disasters brought in disaster researchers from all over the world, including Latin America. In that event, we discovered that there were Latin American researchers from many disciplines doing a great deal of research in many countries in the region, some of whom were anthropologists, including the editor of this volume. For many, it was a watershed moment. While it was not yet recognized as such, it was clear that a social science disaster research tradition in Latin America was in the making. It was a little like discovering that one actually was no longer an orphan but had a lively and diverse kin group.

And within slightly more than a year later, in August of 1992, as noted in several of the chapters in this collection, a core group of people who attended the RGS and UCLA conferences convened in Puerto Limon, Costa Rica, and founded the multi-disciplinary La Red de Estudios Sociales en Prevención de Desastres en América Latina. Although the history of La Red is not the history of disaster research in Latin America, no history of the field can be written without acknowledging its role in the development of social research on disasters not only in Latin America but in the world. However, today the field is far wider and larger than La Red ever was or is. Nevertheless, La Red has influenced theory, methodology, and policy not only in Latin America but in the entire world. Anthropologists were members of the core group and key contributors over the years to its many activities and achievements.

In that context, disaster research in Latin America, like disaster research globally, is multi- and interdisciplinary. Anthropology has been an integral part of that multi/interdisciplinary tradition since its start. When a disaster occurs, every aspect of social and material life is affected, but a disaster is more than the sum of the different kinds of losses and damages it imposes. While every discipline over the entire spectrum of the social sciences has significant contributions to make in advancing the field, the "totalizing" nature of a disaster, impacting virtually every dimension of life, gives the holistic perspective of anthropology special salience in the study of disasters.

Nonetheless, for a long time the contributions of Latin American researchers were not accorded much recognition in the field of disaster research. To some degree this was due to the "unusual" predilection of these researchers to publish their findings and perspectives in their own languages, for their own readers. Moreover, while many of them were and are academics, they were really out to change public policy in their own nations, shifting the focus of disaster management from the purely reactive mode of emergency response and reconstruction to root cause analysis, with the ultimate goal of making disaster risk reduction a goal of national development policy. Writing in English, while perhaps expanding their readership, would have done little to advance that goal. However, their efforts are bearing fruit. Today for example, in Peru there has been significant legislative progress toward integrating disaster risk reduction into the development portfolio with the passage of Law 29644, which created the National System of Disaster Risk Management (SINAGERD) in May of 2011. In September of 2018, Mexico also passed a second version of the 2012 Law of Civil Protection, which similarly prohibits the construction of risk. While deep inroads into long-standing patterns of risk creation have yet to be achieved, these laws are an indisputable result of the efforts of the disaster research community of both nations and the continent as a whole to integrate disaster risk reduction into national development policy.

Over the last 40 years, the social scientific disaster research community in Latin America, as in the whole world, has grown in volume and stature. In effect, disasters in the industrial nations of the global north are no longer the sole focus of attention, as they were prior to the 1970s. Indeed, while research may still be site specific, the focus has actually shifted to a global perspective on risk and disaster events and processes. The linkage between the failures of development and the incidence of disasters, including the attention paid to small and medium-sized events, as first conceived by La Red in the nations of the developing world, now has the full attention of the field. Indeed, perspectives on the social construction of risk, vulnerability, and disaster gained greater attention in the global south, particularly in Latin America, prior to their widespread acceptance in the global north.

As the chapters in this book attest, anthropological researchers in Latin America address the full spectrum of issues that a disaster invokes, some with a specific focus but always with the whole in mind. Nevertheless, the paths of development of Latin American anthropology as well as disaster research have differed by country. These variations reflect not only the variety of hazards experienced according to geography but also the different academic traditions in anthropology, and most especially the political and governance structures and institutions in each nation. Anthropology as a discipline has had a variety of relationships and interactions with national governments, some linked with international power relations of varying description, as noted in this book. For example, the longstanding political involvement of anthropology in Mexican national policy domains, almost unique in the world, has provided fertile ground for the development of an anthropological study of disasters. In some nations, such as Brazil, a myth that there were no natural hazards in that nation, despite a deep and longstanding experience with drought in the northeast, is seen to have inhibited the formation of a social scientific disaster research community. In others, the association of anthropology with movements of liberation, as, for example, in Guatemala, resulted in government persecution, exile, and death of Guatemalan anthropologists working in disaster recovery after the earthquake of 1974. In sum, each nation of Latin America has come to the encounter between natural hazards and society that drives the social construction of risk and disasters from its own cultural and political economic context and history, as well as its own academic traditions and theoretical frames, but each has contributed to the advancement of the holistic anthropological understanding of these complex phenomena for the world.

Anthony Oliver-Smith
Gainesville, Florida
July 2019

Reference

Ortiz, F. (1947) *El huracán: su mitología y sus símbolos*. Mexico Fondo de Cultura Económica.

Acknowledgments

To Anthony Oliver-Smith, for his teachings, which are evident in this book, and for how much Anthropology of Disasters in general, and in Latin America in particular, owes him.

To Ilan Kelman for his support, and for encouraging us to include this book in the Series "Routledge Studies in Hazards, Disaster Risk and Climate Change."

To the three anonymous reviewers of the original proposal of this book, whose comments encouraged us authors to do our best to make this work a remarkable one.

To LA RED, for introducing a new way of understanding disasters and risk through the lenses of Latin American reality.

Introduction

Anthropologists studying disasters in Latin America: why, when, how?

Virginia García-Acosta

One of the main objectives of this book is to show that anthropological production in the study of risk and disasters is quite relevant and has been instrumental in discussions and progress made in Disaster Risk Reduction. Among the contributions anthropology has made to this field are a holistic perspective, the combination of research and practice, as well as an acknowledgement that culture is a totality and that disasters and risk constitute processes that are historically built.

After some pioneering studies in the 1950s, the first anthropological studies on the field began in the 1970s in Europe, in conjunction with the occurrence of some serious disasters. The foray into Latin America among social sciences in general, and as an Anthropology of Disasters began in the 1980s, after several disasters associated to natural hazards happened in the region. Since then relevant research has been done. However, the Latin American dialog around these issues has been little known throughout the world, mainly because the vast majority of the publications is in Spanish. There has been a lot of work, however, which has produced a very important qualitative leap from the decade of the nineties of the 20th century on.

Anthropology of Disasters in Latin America: State of the Art, within the remarkable Routledge Series "Studies in Hazards, Disaster Risk and Climate Change" offers a wide panorama on the subject. This volume brings together a number of experts that show the progress from national and regional perspectives, addressing the birth, evolution, and state of the art of the Anthropology of Disasters in Latin America. It aims to provide a contribution to knowledge coming from what is called the "global south", not understanding it as a geographical notion but as an analytically conceptual one, a critical expression that focuses, in a broad sense, on global differences. I start from the premise that although modes of thought and ideologies that spread from Europe, and afterwards from the United States, to the rest of the world still claim universal applicability, "social sciences do not exist in a cultural vacuum, but develop in specific social and cultural contexts" that differ from the north and the west to the south and the east (Vessuri & Bueno, 2016: pp. 161, 164). I prefer to frame Latin America and the study of disasters, the subject of this book, as part of this global south instead of talking about "third world", "underdeveloped", or "developing" countries. Developing countries? It has been widely recognized that disasters are not only unresolved problems of development, as several specialists have stated (Cuny, 1983; Wijkman & Timberlake, 1984), but problems precisely exacerbated by the development models imposed by levied forms of economic growth, by the methods of accumulation adopted, and by the patterns of settlement and territorial occupation that this development has forced particularly in the global south countries. Disasters are endogenous indicators of processes derived precisely from the models of development and economic growth adopted (Maskrey & Lavell, 2013). They are, as Rogelio Altez mentions in his Venezuela chapter, "critical windows" that allows us to observe the underlying processes and not only the event.

We know that, for decades, different disciplines among social sciences have been interested in the theoretical and methodological study of risk and disasters, and that progress to date shows that interdisciplinary dialog is required. No doubt, there is a lot about geography or sociology of risk and disasters. A lot that is very good indeed. Together with anthropologists, geographers such as Kenneth Hewitt, Ben Wisner, and Phil O'Keefe, with experience in the now called global south, "criticized the essentially passive role prior investigators had assigned to society in risk etiology and the scant attention paid to local, national and international factors in creating or exacerbating both risk and impact" (Oliver-Smith, 2015: p. 547). However, I decided to focus in this book on the Anthropology of Disasters, assuming the risk that this entails. The risk of doing so is high, coming from a discipline and even acknowledging its close bonds with others such as geography and sociology. The risk of doing so reflects the discussion of Immanuel Wallerstein in his article entitled "Anthropology, Sociology, and Other Dubious Disciplines", presented at the Sidney W. Mintz Lecture in 2002. I accept, with him, that disciplines, anthropology among them, "are simultaneously three things": intellectual categories, institutional structures, and cultures (Wallerstein, 2003: p. 453).

Anthropology of Disasters is, as well, those "three things". It has its specificities depending on the context which, I am sure, will be identified throughout the reading of this book that offers a look from a lens focused on Latin America, and addressing nine of its countries/regions: Argentina, Brazil, Central America, Colombia, Ecuador, Mexico, Peru, Uruguay, and Venezuela. I apologize for those Latin American countries and regions not included in this book. Different reasons explain, although they do not justify, their absence.

None of the nine chapters has been published before in the way they are presented in this book. The subjects in this introduction are the chapters themselves, not the authors. I decided on this format to help the reader identify in a better way the elements that distinguish one from another country or region. That does not mean that we do not recognize the effort that each of the 11 authors made in order to search, prepare, analyze and write several drafts of their chapters. I am deeply grateful to them, and the anthropological community will surely recognize them as well:

ANA MARÍA MURGIDA AND JUAN CARLOS RADOVICH: Argentina chapter.
RENZO TADDEI: Brazil chapter.

ROBERTO E. BARRIOS AND CARLOS BATRES: Central America chapter.
ALEJANDRO CAMARGO: Colombia chapter.
A.J. FAAS: Ecuador chapter.
VIRGINIA GARCÍA-ACOSTA: Mexico chapter.
FERNANDO BRAVO ALARCÓN: Peru chapter.
JAVIER TAKS: Uruguay chapter.
ROGELIO ALTEZ: Venezuela chapter.

In terms of the way that present the chapters, I should say the following. Although the authors have common concepts, and even being trained all as professional

Map 0.1 Countries and regions included in the chapters

anthropologists, the vocabulary may vary; as Wisner et al. (2015: v. III) asserts, "much 'hard talk' is required among discipline-based team members before a common vocabulary is achieved", even more so if we are talking in this book about what I have dared to describe as an "adjectival discipline": Anthropology of Disasters.

This introduction is divided into four sections. It begins with the pioneers of Anthropology of Disasters and then presents the review studies that have been carried out at different times during the 20th and 21st centuries. The introduction then continues telling when and how the foray into Latin America took place, and gives a brief account of some of the contributions, as well as some convergences between the nine chapters that comprise it. Finally, I offer some brief reflections on what several authors of the book consider a promising future.

Before fully undertaking the aforementioned sections, a few words about the subtitle of this introduction. It is clear, for the specialized reader, that it is inspired in the one that opens the book *Catastrophe & Culture. The Anthropology of Disaster*, one of those which I have called emblematic publications on the subject. This introduction, written by my dearest colleagues Anthony Oliver-Smith and Susanna Hoffman, has as subtitle the following: "Why Should Anthropologists Study Disasters?" Adapted to the objectives, interests, and contents of this publication, I chose to subtitle it "Anthropologists studying disasters in Latin America: why, when, how?" I hope the answers come throughout this book.

Pioneers and review studies

The study of risk and disasters, particularly those linked to natural hazards, has attracted the attention of social scientists coming from different disciplines over a century ago. Canadian Sociologist Samuel H. Prince is widely recognized as the real pioneer in the social study of disasters by his *Catastrophe and Social Change*, submitted as a doctoral thesis in Sociology at Columbia University in 1920, analyzing the 1917 explosion of a munitions ship in Halifax harbor (Prince, 1920). It was followed by a few social scientists, who did research and published some papers in the next three decades, mainly sociologists like Lowell Juilliard Carr (1932) and Peter Sorokin (1942).

Anthropology entered into these issues with greater intensity over the decade of the 1950s, with studies carried out mostly from British anthropologists. This real and more systematic interest of anthropology in the study of disasters from the fifties on sought "to search for cross-site invariances in disaster-related behaviors, and to codify these findings or to participate in hazard-control planning" (Torry, 1979b: p. 518).[1]

A series of case studies related to specific events, particularly concerned with social changes, were realized and published in journals as *Oceania, Human Organisation* or *Human Relations*.[2] Among them are those related to typhoons among the Yap in the Western Caroline Islands between November 1947 and January 1948 (Schneider, 1957), the eruption of Mount Lamington, so-called mountain Orokaiva in Papua New Guinea on January and March 1951 (Belshaw, 1951; Keesing, 1952 and later Schwimmer, 1969), or the tornado in Worcester, Massachusetts, in April 1953 (Wallace, 1956).

Among them, the ones carried out by anthropologists Raymond W. Firth and James Spillius, New Zealander and Canadian respectively, after the hurricanes that struck in January 1952 and March 1953 Tikopia one of the Solomon Islands, were recognized as the more detailed among them because of their ethnography and for working at the scene of the disaster.[3] Firth, recognized as a classic anthropology by his 1929 research among the Tikopia published as *We, The Tikopia. Kinship in Primitive Polynesia*,[4] went again to the Polynesian island with Spillius, his very dear research assistant, with the objective of observe, record and analyze social changes taken place throughout those more than 20 years; the result was his study about which the author himself described "strictly speaking" as "a *dual-synchronic*, not a strictly *diachronic* study" (Firth, 1959: p. 22). While accepting that the hurricanes, as well as the consequential drought and famine experienced in Tikopia were crucial elements in the changes suffered by the population, not they attributed the disaster exclusively to the passage of the hurricane.

The study considered as the first one whose central interest is disasters associated with a recurrent natural hazard is Canadian-American anthropologist Anthony F. C. Wallace *Tornado in Worcester*. He created what he called a time-space model of disaster as a type of behavioral event, seeking to offer a general model in order for disasters that could be used to systematically compare and analyze with respect to variation along the dimensions of time and space. The Worcester event was "a proper place to apply the early formulations of the model", (Wallace, 1956: p. 1) which includes six essential elements, each of which has devoted to it a chapter of the study: steady state (pre-tornado), warning, impact, isolation, rescue, rehabilitation, irreversible change, and special topics. In this last is included the "Disaster Syndrome" and the "Cornucopia Theory", always combining ethnographic data with its analysis. I agree that Wallace's contributions to anthropological research, besides the Worcester tornado study, linking the issues of disaster, cultural crises, response, and social change constitutes a major contribution to social scientific theory of rapid social change during the rest of the 20th century (Oliver-Smith, 1995: p. 57, 1996: p. 320). The dedication that Oliver-Smith and Susanna Hoffman gave him in the now classic *The Angry Earth* recognizing his work as "*a* Pioneer in the Anthropology of Disaster" is understandable and shared.

The important momentum in the 1950s was not sustained in the following years, until the late 1970s, when there was a rebound with "published research conducted predominantly on herding and mixed- farming populations incapacitated by recent Sahelian and East African droughts", says Torry (1979a: p. 518), recognizing that drought was the most frequently described hazard, specifically among African pastoral peoples.[5] Since then remained some continuity, although with changing approaches. It was since then that was identified, perhaps for the first time, an Anthropology of Disasters. As it is evident by the aforementioned and the references cited, the contributions of two American anthropologists in those years were decisive: Anthony Oliver-Smith, from the University of Florida, Gainesville, and William I. Torry, from the University of West Virginia. Both published important papers at the end of the seventies (Oliver-Smith, 1977a, 1977b, 1979a, 1979b, 1979c; Torry, 1978, 1979a, 1979b). It is precisely these two anthropologists to whom we owe the first two studies of review on the state of the art of the Anthropology of Disasters. Torry's "Anthropological Studies in Hazardous Environments: Past Trends and New Horizons" (1979a), written in 1978 and published in *Current Anthropology* one year later, and Oliver-Smith's "Disaster Context and Causation: An Overview of Changing Perspectives in Disaster Research" (1986b). Both I have addressed before extensively.

The first one is an overview in two directions: what approaches were in the moment on disaster studies, mainly from an anthropological perspective, and which were the research horizons. The second one constitutes the Introduction of the first compilation of articles written by anthropologists in different parts of the world: Latin America, Alaska, Israel, Bangladesh, and Africa. Oliver-Smith prepared this overview, a few years after Torry's, addressing not only a synopsis of the development of disaster studies, but as well questioning the definitions of the main concepts among them and identifying clearly that what was increasing were not the hazards but the vulnerabilities and the risk in face of them, which led him to one of the three conclusions he addresses:

> the impact of human systems on the environment and the increase of human populations in disaster prone regions have increased the risk and scale impact of natural phenomena. In effect, the environment has not become more hazardous. Through human intervention, vulnerability to natural hazard in certain world regions has been increased.
>
> (Oliver-Smith, 1986b: p. 7)

The alternative approach had already positioned itself among the anthropologists.

Oliver-Smith's 1996 review was, up to that time, the most comprehensive ever done. It was published first in Spanish in LA RED's journal *Desastres & Sociedad* (*Disasters & Society*) and one year later in the *Annual Review of Anthropology* (Oliver-Smith, 1995, 1996). In it he analyzes in depth the three general approaches that anthropology had developed until then: the behavioral response one, the social change one and the political-environmental one, a separation he considered artificial, as the three of them "address issues that are related causally, developmentally and conceptually", and insists that although a multidisciplinary approach is needed, anthropology is the one which addresses the issue holistically (Oliver-Smith, 1996: pp. 305, 322).

Subsequent review studies, none of them so complete as the latter, were published again until the 21st century. Two of them constitute, as others, introductions to special editions of two journals: *Human Organization*, which has included reflections in the field almost since it was born in 1941, and *Iberoamericana-Nordic Journal of Latin American and Caribbean Studies*, whose last transformation to what it is now began in de 1990s. The third and fourth ones, more retrospective, were published in *The International Encyclopedia of the Social and Behavioral Sciences* in 2015 and in *The International Encyclopedia of Anthropology* in 2018. Before these four reviews, without antecedent or continuity on the part of its author, Doug Henry published a review in 2005, which is interesting as it introduces some issues scarcely appreciated by specialists until then like coping mechanisms and capacity of recovery; although Latin American disaster anthropology had already important advances, only Oliver-Smith's work on Peru mentions it (Henry, 2005).

A.J. Faas and Roberto E. Barrios (2015), American and Guatemalan-American anthropologists, respectively, present a new retrospective almost 30 years after the 1986 one aforementioned. The authors talk about advancements, historical antecedents, methodological approaches, and theoretical learnings that they identify as the "today's diversified field of disaster anthropology". It draws on seven articles coming from different regions of the world, addressing diverse natural hazards and bringing up new ideas and concepts as that of "procedural vulnerability" in the case of the 2009 Taiwan typhoon (Hsu et al., 2015).[6]

Swedish anthropologist Susann Baez Ullberg's (2017) overview is rather shorter and focuses on disasters and crises or "critical events" also from the perpective of social anthropology. Along with Argentinian anthropologist Sergio Visacovsky in his chapter included in this *Iberoamericana* Special Collection, clarify why they talk about disasters and crises, two concepts they consider close connected, with arguments regarding "what we call 'economic crises' have disastrous aspects. In turn, since disaster breaks the temporal continuity, the current time is perceived as 'frozen' and the future cannot be imagined, this implies installing a time of crisis" (Visacovsky, 2017: p. 7).[7]

The entries in the two encyclopedias mentioned before are concise, as those entries should be (Oliver-Smith, 2015; García-Acosta, 2018). Both recognize disasters as highly complex and multidimensional processes, which materialize in time and space in a specific event and that are better understood by the holistic perspective of anthropology. They refer to the field from a global point of view highlighting, with different emphasis, the importance of cultural ecology and political ecology in the anthropological research, and equally recognize the paradigm shift of the 1980s, which really started in the mid-1970s. Oliver-Smith's entry concentrates on how notions grew and changed, discusses old concepts, introduces new ones, presents innovative paths like the idea of mutuality between society and nature that resides at the core of any disaster, and which is clearly expressed in the challenges presented by global climate change, which represents one of the gaps between disaster research and practice. García-Acosta's text is a narrative of how the field was born, its evolution, the contributions of the discipline, the state of the art in some regions of the world, and its representation in institutions such as meetings and journals.

This book is centered in Latin America. So I checked the aforementioned texts looking for the following: what do they include about this region? Was the Anthropology of Disasters present in it and since when? As the overviews cover from the late seventies to today another question was this: how has been Latin America present in the Anthropology of Disasters in the 1970s, 1980s, and in the second decade of the 21st century?

In Torry's reviews (1979a, 1979b) Latin America appears only related to case studies conducted in Peru and Mexico, carried out by Latin Americanists rather than by anthropologists from the same region. In the first case, he mentions Oliver-Smith, thanking him for his comments to the text, and afterwards mentioning the "1970 Peruvian Cataclysm" and the authors that until then had studied that event besides Oliver-Smith: P.L. Doughty, S.W. Dudasik, B. Bode, and J. Osterling. In the Mexican case, the references allude to mainly tectonic natural hazards related to the end of the Maya era; among them are E.W. Mackie, R.J. Sharer, and R.E.W. Adams, as well as the archaeologist P.D. Sheets.

Oliver-Smith in his 1986 Introduction to a special issue of *Studies in Third World Societies* mentions three Latin American cases studied by anthropologists: the Peruvian earthquake in 1970 (B. Bode), 1974 Hurricane Fifi in Honduras (no specific authors), and 1975 Guatemalan earthquake (J.F. Alexander, N.S. Sipe, W. Peacock, and F. Long), The issue includes seven articles, two of which refer to Latin America: S. Robinson et al., "It Shook Again. The Mexico City Earthquake of 1985" and P.L. Doughty "Decades of Disaster: Promise and Performance in the Callejon de Huaylas, Peru".

Reviews coming from the 21st century show a completely different picture. These illustrate how the panorama did change in those three decades, mainly concerning the increasingly stronger inclusion of Latin America in the anthropological research on disasters close to the ending of 20th century.

Faas and Barrios, as editors of the special issue that was the product of a call for papers for the 2013 Society for Applied Anthropology Annual Meeting motivated by the granting of the Malinowski Award to Anthony Oliver-Smith, introduce seven articles. Only two of them are the product of research in Latin America: Brazilian V. Marchezini's "The Biopolitics of Disaster: Power, Discourses and Practices", and "The Construction of Vulnerability along the Zarumilla River Valley in Prehistory" by Sarah Taylor, addressing Brazil and the Peru-Ecuador border, respectively.

In 2017 the first collection of articles appeared that was dedicated exclusively to the field in the region (Baez Ullberg, 2017). As was to be expected since it is announced in the title ("The Contribution of Anthropology to the Study of Crises and Disasters in Latin America"), the articles concerning case studies included in this *Iberoamericana* special collection are based on empirical information and come from research in and about the region, although addressing only two spaces: Argentina and Brazil. In the first case, D. Zenobi refers to the political dimensions of the 2004 nightclub fire known as the "Tragedy of Cromañón" and Baez Ullberg to the 2003 catastrophic flood occurred in the city of Santa Fe.[8] As far as Brazil is concerned, T. Camargo da Silva focuses on the 1987 Goiânia radiological disaster from the point of view of young people's narratives expressed almost 20 years after.

The contributions of the Red de Estudios Sociales en Prevención de Desastres en América Latina (Network for Social Research on Disaster Prevention in Latin America) known as La RED, is recognized by all the aforementioned reviews as remarkable. First, because it did really introduce the discussion in the region, and second because it contributed to the expansion of the originally known as the "alternative approach" and later on the "vulnerability approach. *The International*

Encyclopedia of Anthropology entry focuses more in the region, highlighting research done in a marriage of sorts between anthropology and history in Mexico and Venezuela, and recognizing that in other Latin American countries, disasters anthropology was still a subject with an incipient presence, although not entirely absent from anthropological production.

I hope that the nine chapters of this book can change or at least moderate this perspective.[9] Now let us go and examine how this incursion in Latin America began.

The foray into Latin America

In the last two decades of the 20th century, interest in the field began its foray into Latin American areas dedicated to research and teaching in anthropology. If the turning point for a "formal" beginning of an Anthropology of Disasters are the 1970s with the reflections and contributions of Oliver-Smith and Torry, the watershed in Latin America began with the concatenation of a series of disasters associated with natural hazards, geological or hydrometeorological, in the region: for example the 1970 earthquake and subsequent landslide in the Callejón de Hua-ylas in Peru, the Chichonal Volcano eruption in 1982 and its multidimensional consequences among Zoque Indigenous people in Mexico, or that of the Nevado del Ruiz volcano in Colombia in 1985 that led to the Armero tragedy, together with the 1982–1983 El Niño Southern Oscillation's effects and impacts mainly in Ecuador and Peru, or the Mexico City earthquake in 1985.

We must remember that there is an important antecedent that goes back as the 1940s, identified with a pioneer, the Cuban anthropologist Fernando Ortiz and his book *El huracán: su mitología y sus símbolos* (*The hurricane: its mythology and its symbols*) (Ortiz, 1947). To it must be added, although several decades after Ortiz, a couple of products of research, not identified with a specific field, not even referring to disaster or risk, but that in the Mexican case have been remarka-ble contributions to the development of the field. The first one is a research project carried out by a group of Mexican anthropologists after the eruption of the Chi-chonal volcano in Chiapas, with previous experience in the area, trying to identify comparatively the social, cultural and economic effects (Báez-Jorge et al., 1985). The second one is Canadian anthropologist Herman Konrad's analysis of the role of tropical storms between pre-Hispanic and contemporary Maya, considering a phenomenon overlooked although deeply rooted in the Maya society and culture (Konrad, 1985).[10]

In the same way that Wallace's *Tornado in Worcester* was the first book entirely dedicated to the Anthropology of Disasters published in 1956, the field had to wait three decades to receive Oliver-Smith's *The Martyred City: Death and Rebirth in the Andes*, published in 1986 as the first book entirely dedicated to the Anthropol-ogy of Disasters in Latin America (Oliver-Smith, 1986a).

In the chapters that constitute the core of the present book, we can find a number of important studies which, as in the Mexican and Peruvian examples, preceded the definitive incursion of Anthropology of Disasters in Latin America, and that until now had not been systematized and made known jointly.[11] Seasonal floods in Argentina, floods related to changes in land-use and their effects were studied since the 1970s, as well as the social and cultural impacts of forced displacement related to hydroelectric dams in the same decade in the region (Argentina, Brazil, Mexico). The 1976 Guatemala earthquake instigated anthropological research in the area. In pre-Columbian Ecuador, Mexico and Peru, archaeological research on historical disasters goes farther than that. Spanish chroniclers, old newspapers, and 19th-century scientists have inherited important information about earthquakes, volcanic eruptions, landslides, or the presence of hurricanes throughout the Hispanic Monarchy.

In fact, the region has been identified, for a long time, as a territory where large-magnitude events take place. Avalanches, earthquakes, droughts, El Niño, floods, frosts, hailstorms, landslides, mudslides, snowfalls, tornados, volcanic eruptions, windstorms, and even those with local names as *huaycos* in Peru[12] have been part of the daily lives of those who have lived there for centuries. Some of them are indeed identified as "traditional" or recurrent hazards because of their regularity. Faced with this evidence the paradox is this: why in several regions or countries has anthropology not attended to this issue? Why is it that several of the chapters of this book refer to what they call explicitly or implicitly "invisible disasters", which appear expressly in the chapters about Brazil, Central America, Colombia, and Peru? The chapter about Central America even identifies this contradiction as a "key tension".

In certain Latin American countries, the limited development of the Anthropology of Disasters has to be understood aside from the narrow development of anthropology in general. However, what happens to countries where it is recognized they have a well-developed and internationally recognized anthropology, which has even been qualified as a "vibrant anthropological community" like Brazil? Or featuring dozens of graduate and postgraduate programs in the discipline, distributed among various university departments of anthropology or social sciences? Or with a high quality anthropological research and a good amount of well recognized anthropological journals?

The causes of this contradiction between being fertile spaces for disaster research and the scant attention anthropologists have devoted to it are explored (Brazil, Colombia, and Peru chapters). The reader of this book will find that there is no single answer, although there are several coincidences. In some chapters, the explanations are explored thoroughly. One of the answers I considered when I started thinking about this book appears among them: maybe what has happened is that anthropologists had studied them, that disasters have in fact been part of the ethnographies of Latin American anthropologists, but what was missing was to track such efforts and put them together. If this is true now, in the next version of the recently published *The International Encyclopedia of Anthropology*, which includes the development of the discipline in five Latin American countries (Argentina, Brazil, Chile, Colombia, and Mexico), Anthropology of Disasters will be mentioned as one of the fields cultivated in the region.[13]

A key role in the start of a critical discussion from the social sciences to the study of disasters in Latin America was the creation of the aforementioned Network for

Social Research on Disaster Prevention in Latin America, LA RED. Created in 1992, its products "have been influential to the point that today the contents of many of the new laws and public policies developed in Central America and the Andean countries from the mid-1990s onwards have reflected its basic concepts" (Lavell, 2017: p. 15). LA RED has included since the beginning Latin American and Latin Americanist anthropologists among its members, as well as products of their research in many of its publications. They had considerable influence in the change of paradigms that LA RED launched. The relevance of LA RED is recognized in almost all the chapters of this book, commenting on the way it filled the gap in the region with an academic and a political agenda, its contributions, making them available in Spanish, and even mentioning specific names.[14] The publications of LA RED are widely cited in the bibliographies of each chapter.

The first product related to Anthropology of Disasters that appeared as a result of the projects, seminars, and conferences that LA RED organized and/or sponsored were the first two volumes of *Historia y Desastres en América Latina* (*History and Disasters in Latin America*, 1997, 1997), to which a decade later a third volume was added (2008). They include articles from archaeology, history, and social anthropology on the great majority of countries in the region.[15] These studies were added to what each country had been producing independently, and to what is chronicled in the chapters of this book.

The production of studies and publications on Anthropology of Disasters focused on Latin America was gradually expanding in different ways. For example, several compilations on the subject that appeared since the end of the 1980s gradually began including, increasingly, research cases coming from Latin America. Some examples of publications that start in the 1980s and cover three decades more are as follows:

a   *Natural Disasters and Cultural Responses* (Oliver-Smith, 1986b) include Mexico and Peru.

b   *The Angry Earth: Disaster in Anthropological Perspective* (Oliver-Smith & Hoffman, 1999): Peru.

c   *Constructing Risk, Threat & Catastrophe. Anthropological Perspectives* (Giordano & Boscoboinik, 2002): Costa Rica, Honduras and Mexico.

d   *Catastrophe & Culture. The Anthropology of Disaster* (Hoffman & Oliver-Smith, 2002): Mexico and Peru.

e   "Applied Anthropology of Risk, Hazards, and Disasters" (Faas & Barrios, 2015): Brazil, Ecuador/Peru border.

f   "La Contribución de la Antropología al Estudio de Crisis y Desastres en América Latina" ("The Contribution of Anthropology to the Study of Crises and Disasters in Latin America") (Baez Ullberg, 2017): Argentina and Brazil.

g   *Les Catastrophes et l'interdisciplinarité: dialogues, regards croisés, pratiques* (*Disasters and Interdisciplinarity: Dialogues, Crossed Glances, Practices*) (García-Acosta & Musset, 2017): Haiti and Mexico.

h   *Antropología, Historia y Vulnerabilidad. Miradas diversas desde América Latina* (*Anthropology, History and Vulnerability. Different Perspectives from*

Latin America) (Altez & Campos, 2018): Brazil, Guatemala, Mexico and Venezuela.

i   *Disasters in Popular Cultures* (Gugg et al., 2019): Latin America.

Over the last ten years, we have four excellent books that have continued the tradition, which I hope will continue, to publish books in which the theme is the Anthropology of Disasters in Latin America focused in one country, in one region. They are the following: Sandrine Revet, *Anthropologie d'une catastrophe. Les coulées de boue de 1999 au Venezuela* (*Anthropology of a disaster. The mudslides of 1999 in Venezuela*), 2007; Susann Ullberg, *Watermarks. Urban flooding and Memoryscape in Argentina*, 2013; Julie Hermesse, *De l'ouragan à la catastrophe au Guatemala. Nourrir les montagnes* (*From the hurricane to the disaster in Guatemala. Nourish the mountains*), 2016, and Roberto E. Barrios, *Governing Affect. Neoliberalism and Disaster Reconstruction*, 2017.

And counting . . .

In the first decade of the 21st century, the subject of Anthropology of Disasters in Latin America began to show up increasingly in general meetings of anthropologists at national or international levels.[16] Associations, working groups, or even conferences who already had as a central theme the Anthropology of Disasters were gradually emerging.[17] As part of their agendas, discussion, and developments carried out in the Latin American region were introduced gradually. Among them the TIG (Risk and Disaster Topical Interest Group) of the SfAA (Society for Applied Anthropology), created in 2013 deserves special attention.[18]

Two of these meetings constituted the main background of the nine chapters of this book. The worktable in the framework of the ALA Conference in Bogota in 2017 and the Symposium developed at IUAES Conference celebrated in Florianopolis 2018, in both cases under the title of Anthropology of Disasters in Latin America: State of the Art and having Gonzalo Díaz-Crovetto and Virginia García-Acosta as conveners.[19] One of the objectives on which we insisted from the beginning was to avoid in this, and other fields of anthropological knowledge that continue to dominate, what we have called "academic stiff neck", which tends to look and value exclusively the theoretical paradigms produced in the global north of the world. A favorable situation and an opportunity to reflect on and disseminate what in this regard exists, which is being done and has to be done in the global south, is the approval in 2018 of the IUAES Commission on Risk and Disaster Anthropology.[20]

The structure of the book

All the chapters of this book refer to a specific country, highlighting regions, entities, or specific spaces. There is an exception in the case of Central America, which is approached as a region and basically includes two countries: Guatemala and Honduras, and to a lesser extent El Salvador, Nicaragua and Costa Rica.

Each of the chapters offers a general panorama on the evolution of anthropology in the country or the region, as well as the emergence and evolution of what could be called in generic terms anthropological studies on risks and disasters, each presenting the current state on the Anthropology of Disasters. These contents were suggested at the beginning of this project, and each author developed them according to his/her own format, which involves his/her background, his/her narrative, his/her vocabulary and ways of transmitting them.

These chapters emphasize the importance of doing situated ethnography, of using traditional anthropological techniques as participant observation and intensive fieldwork in the studied area. This allows for really knowing and understanding the context where disastrous processes happen.[21] It is one of the clues to risk assessment, as it helps to identify not only how risks are socially constructed in face of natural hazards in specific contexts, but also to recognize cultural perceptions of risk.

Some had already worked the theme in their countries, others had to start real surveys adopting broadly similar methodologies. For the chapter on Colombia a number of conversations with anthropologists who have either studied or being in the aftermath of a disaster event were developed. In the case of the Uruguay chapter, a search in academic journals was done, as well as interviews with colleagues in University's Anthropology and Archaeology departments. The rest of the countries, where Anthropology of Disasters was considered as "invisible", were the product of similar exercises.

No doubt the chapters respond to what has been called "ethnographic curiosity" accompanied by a suitable methodology. With ethnographic curiosity and, as well, giving what in the chapter on Ecuador is seen as a closer reading of anthropological studies which leads to find invaluable information coming from local memories and narratives.

To conclude this section, I would like to refer to two elements that were included in all chapters. On one side, and in response to a request I made, in the texts and also in the references are mentioned and even discussed those BA, Master, or Doctorate theses related to an Anthropology of Disasters in the corresponding country. The total number is striking as it reaches almost half a hundred, even considering that perhaps there may exist an underreporting.[22] On the other side, each the nine chapters includes a map in which are represented exclusively the spaces referred to in the corresponding text, either because in them there were carried out anthropological case studies or because there happened specific events worthy of highlight. So the title thereof for all of the maps is preceded by the name of the country or region: "Case studies and main areas mentioned".[23]

Contents, outcomes, and debate

This book offers different approaches and theoretical perspectives, many of which even we did not imagine could appear. Some confluences among the nine chapters will be presented in this section in the hopes of instigating an occasion to reflect and debate about the current state and future development of Anthropology of Disasters in general and in Latin America in particular.

First of all, there is the assumption that disasters are not natural, which data presented in all the chapters confirm, recognizing that vulnerability is in the core of disaster occurrence. Second, there is the recognition that disasters are processes, which implies that the unavoidable study, analysis, and understanding in a historical perspective, is present in most of them. In the Mexico's, Venezuela's, and Central America's chapters, the point is more explicitly stated.

The importance of the context in which the process of construction of the disaster takes place, as well as the moment in which it becomes a reality (the event), is evident in all the chapters. The case of indigenous communities living at risk is part of this problem, specifically in what is seen as lack of access to safe living spaces, linked to negligence and mismanagement. Central America's chapter refers to the case of the El Cambray 2 community, south of Guatemala City, a disaster clearly linked to unplanned urban development processes

Inter or multidisciplinarity is another issue that several chapters address, referring to two spheres of knowledge: between social sciences and natural or physical sciences, or among different social sciences in relation to each other. In studying disasters associated with natural hazards, the discussions with climatologists, geomorphologists or seismologists, and volcanologists, with whom many of us have worked, is not only necessary but also required.

In the chapters of this book, the role of geographers is also highlighted.[24] The Argentina chapter recognizes that the participation of geographers "in a region that is exposed to a variety of hydro-climatic and seismic risks [has provided] a key reference on the topic of environment and desertification in Latin America and the Caribbean". The chapter on Ecuador even includes a section entitled "Critical Geography, Environmental Studies, and Risk Management", and concludes with an injunction for greater interdisciplinary dialog. In some cases, and mainly in the last decade and a half, it is recognized that extreme weather events, mainly drought, have provoked interdisciplinary and multidisciplinary discussions under the umbrella of the debate on climate change, as is addressed in the Uruguay chapter.

Multidisciplinarity between social sciences and other disciplines outside them, including specialists dealing with climate change, has been of great benefit to Colombian anthropologists, who are developing important discussions which constitute a fertile terrain for the theorization of "disastrous landscapes".

'Disastrous landscapes' is not only the title and a whole section of the chapter on Colombia, it even constitutes the axis around which it develops, starting from the premise that the "disaster is, primarily, a material phenomenon with an obvious and enduring imprint on the landscape. . . . [So] the materiality of a disastrous landscape significantly affects the ways in which anthropologists understand and approach their research". Uruguay's chapter revisits the concept since considers it useful "in orienting anthropological research towards a more direct engagement with natural sciences knowledge". The discussion around the concept of disastrous landscapes is really interesting and innovative in the field of disasters, which invites us to discuss and, even, to dissent.

The concept of uncertainty is another specific conjunction among the chapters. The Argentina chapter refers to uncertainty when rural or urban populations are affected by natural phenomena, which happen recurrently in all Latin American regions. A "state of uncertainty" is revealed as the product of a combination of natural and technological factors, social organization, adaptive strategies, and models of development, which can be reduced by the inclusion of diverse actors and knowledges through planning and management processes. Uruguay's chapter, referencing Argentina's one as well as Taddeis' article "The politics of uncertainty",[25] notes that one of the main findings was the differences between climate scientists, end-users and nonexperts (like decision-makers, producers and agricultural technicians, or journalists) in their understanding of climate; it can be either a neutral object, something subjective part of embodied memories, or a "byproduct of poor science and limited knowledge", which is incapable of "recogniz[ing] uncertainty". Colombia's chapter only mentions uncertainty once, related to landslides that "paved the way for the implementation of a new regime of risk governance", creating uncertainty among people.

In the three chapters, the concept of uncertainty is important, referring to an imprecision when looking after the frontier between knowledge and certainty; thus "uncertainty can be defined as a gap in the intelligibility of risk, disasters and their effects among territorialized agents which limits the capacity to take right decisions".

Post-normal science or post-normal perspectivism is an issue that is linked to uncertainty, and proposes a dialog that has to be interdisciplinary and, as well, has to interact with a participative methodology that includes scientists and decision-makers. Brazil's analysis goes further and considers disasters as a "form of post-normal perspectivism of contemporaneity" that have definitely to be explored more thoroughly.[26]

Although there are convergences and concurrences among the chapters, there are as well varied perspectives and interpretations, which confirms that research on disaster and risk even from an anthropologian perspective is not unified epistemologically and neither does it represent a single theoretical scope. These chapters on nine Latin American countries include a considerable diversity of approaches and concepts, which reflects a similar variety in the ways of understanding societies and their processes.

Final remarks: promising future

The nine chapters that compose this book make it clear that Anthropology of Disasters in this American region, although in some cases still incipient, is on the right path to build its own anthropologies on a regional scale, responding to their contexts, characteristics, and specificities as "a many-faceted Latin American anthropology of the global South [. . .] to make diversity, as a principle of anthropology (the science of sociocultural diversity), more visible on a planetary scale" (Krotz, 2018)

After the richness and diversity offered by the chapters of this book in terms of concepts and frameworks, the need for encounters and discussions that promise to be very fruitful between different specialties, mainly between subspecialties of anthropology, is inevitable. Among them are Ecological Anthropology, Environmental Anthropology, Anthropology of Landscape (where concepts like disastrous landscapes come from), and, of course, Anthropology of Disasters. To them we could add others that offer possibilities of fruitful dialogs, like Anthropology and Climate Change.[27]

Considering the recently published *International Encyclopedia of Anthropology* an updated compendium of anthropological research, I checked those subfields of anthropology that refer to ecology, environment, landscape, climate change and so and found that Ecological Anthropology and Environmental Anthropology are included.[28] Nevertheless, only the latter refers expressly to our issue, recognizing the re-study in Tikopia by Raymond Firth as one of the insightful studies with which "anthropology was graced" at the middle of the 20th century, and Anthony Oliver-Smith's *The Martyred City* as another of the pioneers. The authors, Cortesi et al. (2018), even identify that "in disaster research disasters are often politically defined as 'environmental' in order to absolve powerful actors of their responsibility for them".

Much more could be said, critically, about this fragmentation of knowledge that, with different labels, in many cases refers to similar or even identical problems. Without doubt, we are faced with a challenge to explore in the Anthropology of Disasters in general what the gaps are between studies on disaster risk reduction, climate change, climate change adaptation, and sustainable development from an anthropological perspective. What are their connections and differences? One key question is the following: is this fragmentation useful when we are talking about application in such critical issues as disaster risk reduction?

Having as an axis the subject of risk and disasters, we must deepen in the relationship between those concepts and subdisciplines that several chapters of this book relate. Promote and strengthen what is called a "fraternal dialog" between them. Already Oliver-Smith (2017) and Hoffman (2017) advanced this discussion with their chapters recently published in the *Routledge Handbook of Environmental Anthropology*, and I think that discussions in the chapters of *Anthropology of Disasters in Latin America: State of the Art*, specifically those dedicated to Brazil, Colombia, Peru, and Uruguay, will be food for thought in this path.

Notes

1 More detailed information about these pioneers can be found in Oliver-Smith, 1986b; Torry, 1979a, 1979b and in almost all of the Mexican BA, Masters, or PhD thesis on Anthropology of Disasters (see References section in Mexico's chapter in this book).
2 Not yet in *Current Anthropology,* as should be expected given that this remarkable journal has dedicated many pages to studies on Anthropology of Disasters, because it was founded in 1959; nevertheless, in *Current Anthropology* appeared one of the best reviews about these issues in the seventies.
3 In the excellent compilation of "musts" in the study of disasters, published in 2015 under the title of *Disaster Risk*, seven out of the 98 articles included correspond to what one may call pioneering social studies on disaster risk during the first half of the 20th century: after Prince's dissertation (1920), one published in the thirties, another one in the fourties, and three more in the fifties. Among them appear three of the classics already mentioned and published within those four decades: sociologist Carr (1932), and the anthropologists Schneider (1957) and Spillius (1957) whose articles were

published the same year (Wisner et al., 2015). In this compilation, the editors included British anthropologist Audrey I. Richards' chapter 2 from *Hunger and Work in a Savage Tribe: A Functional Study of Nutrition Among the Southern Bantu* published in 1932. Although, it cannot be considered as a pioneer anthropologist of disasters but as Ben Wisner himself explains, Richards' chapter was included as it shows that the ability to cope with shocks is rooted in daily life and practice, as is vulnerability (Wisner personal conversation, 2 July 2019).

4 Published in 1936 with a Preface of Bronislaw Malinowski where he classifies Firth's book "as a model of anthropological research, Both as regards the quality of field-work on which it is based and the theories which are implied in it" (Malinowski, 1936: p. vii). I will always be grateful to my colleague JC Gaillard for giving me an original of this jewel of anthropology.

5 The research done in the 1970s by three remarkable and critical social geographers in those areas has been a landmark for all social scientists in this field: Phil O'Keefe, K. Westgate and, up to now, Ben Wisner. Altogether published as early as 1976 was the unavoidable "Taking the naturalness out of natural disaster" (O'Keefe et al., 1976).

6 It is one of the concepts mentioned in the chapter on Ecuador in this book, although the reference is not the same.

7 This is why his article's title is "When Time Freezes". This is a discussion that goes beyond the purposes of this introduction, as some may think the difference is just a semanthic one, as the author himself recognizes; but Visacovsky's article is revealing in many senses around this and other discussions related to the use of different concepts when addressing similar problematics. It is worth reading it and it is preparatory to further discussions on the subject. I draw attention that the name of one of the newly created networks of anthropologists studying this issues is the European Association of Social Anthropologists (EASA) Disaster and Crisis Anthropology Network (DICAN).

8 These articles, as well as Baez Ullberg Introduction, are revised in the Argentina chapter in this book.

9 In the same way Wisner et al. (2015) did in the four volumes of *Disaster Risk*, R. Altez and V. García-Acosta are preparing the publication of a compilation of what we have called "must-read texts" (*textos imperdibles*) or "major works" in the Anthropology of Disasters and Risk, in Spanish. The aim is to help increase and strengthen that knowledge that is still missing in the Spanish-speaking world, mainly among students.

10 To them refers more explicitly Mexico's chapter in this book.

11 The information listed later can be found with more details, in each of the chapters of this book.

12 Peru's chapter in this book mentions "*huayco* is a Quechua Word used in Peru to mean torrential and fast-flowing streams of turbid water running down from the highest part of mountains dragging stones, brushwood, and other sediments along a course of ravines to reach towns and other human settlements."

13 In the chapter "Anthropology in Mexico", the field is mentioned at the end as "See also: Disasters, Anthropology of", as it is one of the entries of *The International Encyclopedia of Anthropology*.

14 Among them are expressly mentioned in the chapters the following, most of them founders of LA RED: Hilda Herzer (+) in Argentina; Gustavo Wilches-Chaux, Omar Darío Cardona, María Teresa Findji, and Víctor Daniel Bonilla in Colombia; Allan Lavell in Costa Rica; Elizabeth Manzilla, Jesús Manuel Macías and Virginia García-Acosta in Mexico; Andrew Maskrey (LA RED creator and first coordinator), Anthony Oliver-Smith and Eduardo Franco (+) in Peru. In the aforementioned compilation, there appears among the selected original texts by several of these founders of LA RED: Volume I "Big-Picture Views": Lavell, Oliver-Smith & Wilches-Chaux; Volume III "Knowledge and Wisdom": García-Acosta & Maskrey; Volume IV "Having Influence": Oliver-Smith &, Wilches-Chaux (Wisner et al., 2015).

15 Argentina, Bolivia, Brazil, Chile, Colombia, Costa Rica, Ecuador, El Salvador, Mexico, Nicaragua, Panama, Peru, and Puerto Rico.

16 In the following meetings, papers were presented on the subject and on different Latin American countries or regions: ALA (Asociación Latinoamericana de Antropología/ Latin American Anthropological Association) in 2012, 2015, and 2017; COMASE (Congreso Mexicano de Antropología Social y Etnología/Mexican Congress of Social Anthropology and Ethnology) in 2010 and 2012; COMECSO (Comité Mexicano de Ciencias Sociales/Mexican Social Sciences Committee) in 2012; EMBRA (Encuentro Mexicano-Brasileño de Antropólogos/Mexican-Brazilian Meeting of Anthropologists) in 2013, 2015 and 2017; IUAES (International Union of Anthropological and Ethnological Sciences) in 2016 and 2018; RBA (Reunión Brasileña de Antropología/Brazilian Meeting of Anthropology) in 2016.

17 Such is the case of ARCRA (Association pour la Recherche sur les Catastrophes et les Risques en Anthropologie/Association for Research on Disasters and Risk in Anthropology); DICAN/EASA (Disaster and Crisis Anthropology Network/European Association of Social Anthropologists) since its foundation in 2014.

18 TIG/SfAA has organized annual meetings on the subject, in which papers on the following countries have been presented: Albuquerque 2014: Argentina, Bolivia, Brazil, Cuba, Ecuador, Guatemala, Haiti, Mexico, Peru, St. Lucia; Pittsburg 2015: Haiti, Mexico; Vancouver, 2016: Belize, Bolivia, Haiti, Mexico; Santa Fe 2017: Brazil, Costa Rica, El Salvador, Mexico, Peru; Philadelphia 2018: Belize, Bolivia, Cuba, Guatemala, Puerto Rico; Portland 2019: Belize, Bolivia, Ecuador, Haiti, Mexico, Puerto Rico.

19 Almost all the authors who are now in this book already participated in 2017: A. Murgida (Argentina), T. Hanson (Bolivia), R. Taddei (Brazil), R.E. Barrios (Central America), G. Díaz-Crovetto (Chile), A. Camargo (Colombia), A.J. Faas (Ecuador), V. García-Acosta (Mexico), J. Taks (Uruguay), and R. Altez (Venezuela). It was there when, fortunately, Peru was added with F. Bravo. In addition to the aforementioned, in 2018 J.C. Radovich participated, who joined as coauthor of the chapter on Argentina. The same happened in the case of Central America's chapter with Carlos Batres.

20 Its leader is one of the main specialists in the subject: Susanna Hofmann, who invited V. García-Acosta as deputy leader.

21 Cfr. the excellent compilation of texts about fieldwork specifically in Latin America recently published by Argentinian anthropologist Rossana Guber, showing "how social anthropologists, cultural anthropologists and ethnologists do research [with] reflections, experiences and precepts about our ways of thinking, doing and imagine the ethnographic field" (Guber, 2018: p. 45).

22 Most of them have been defended in Mexico (13), Colombia (7), and Venezuela (6).

23 I would like to thank geographer Armando Nava for giving these maps a final and homogenized format. And to Diego Vargas for his assistance in the process of conforming this book.

24 Mexican geographer-anthropologist J.M. Macías, in his chapter appeared in a book on anthropology and interdisciplinarity, where he relates anthropology and geography in disaster research, underlines specifically spatial dimensions, regional considerations, and scales (Macías, 2013).

25 Brazil's chapter does not refer to this article, nor to uncertainty.

26 Delving into this topic, without referring explicitly to the field of disasters, anthropologists Taddei and Hidalgo claim that a post-normal anthropology "allows [us] to refer and to conceptualize situations in which the ethnographic encounter takes place in contexts of real ontological clash, and where the conceptual frameworks that structure the perspective of the ethnographer cannot remain unchanged" (2016: p. 22).

27 It is the title of two very interesting volumes coordinated by Susan Crate and Mark Nutall: *Anthropology & Climate Change. From encounters to actions* (Left Coast Press, 2009) and *Anthropology & Climate Change: From actions to transformations* (Routledge, 2016). Crate is the author of the entry on climate change in *The International Encyclopedia of Anthropology*, and links disasters to displacement.

28 Other entries can be found in that Encyclopedia linked to disasters like the following: climate change, cultural adaptation, cultural ecology, environmental vulnerability and resilience, historical ecology, and political ecology. However, only three cases refer to the issue: climate change (already mentioned), environmental vulnerability and resilience, and political ecology. In the latter it is acknowledged that it "grew out of critical thinking about ecological crises that were not simply 'natural' in their origins and that were experienced by groups of people whose vulnerability to **disasters such as floods, desertification, and landslides** was significantly influenced by their lack of social and economic power" (Campbell, 2018). Use of bold is mine, highlighting that many times, even when recognizing that disasters are not natural, there exists the inertia of identifying or mentioning them as synonymous with natural hazards).

References

Altez, R. & Campos, I. (eds.). (2018) *Antropología, Historia y Vulnerabilidad. Miradas diversas desde América Latina*. Zamora, Mexico, El Colegio de Michoacán.

Báez-Jorge, F., Rivera, A. & Arrieta, P. (1985) *Cuando ardió el cielo y se quemó la tierra. Condiciones socioeconómicas y sanitarias de los pueblos zoques afectados por la erupción del volcán Chichonal*. Mexico, Instituto Nacional Indigenista.

Baez Ullberg, S. (2017) La Contribución de la Antropología al Estudio de Crisis y Desastres en América Latina. *Iberoamericana-Nordic Journal of Latin American and Caribbean Studies*. 46, pp. 1–5.

Belshaw, C. (1951) Social Consequences of the Mount Lamington Eruption. *Oceania*. XXI (4), pp. 241–252.

Campbell, B. (2018) Political Ecology. In: Callan, H. (ed.). *The International Encyclopedia of Anthropology*. New York, John Wiley & Sons, Ltd. DOI:10.1002/9781118924396. wbiea2315.

Carr, J. C. (1932) Disaster and the Sequence-Pattern Concept of Social Change. *American Journal of Sociology*. 38 (2), pp. 207–218.

Cortesi, L., Lennon, M., Hebdon, C., Stoike, J. & Dove, M. R. (2018) Environmental Anthropology. In: Callan, H. (ed.). New York, *The International Encyclopedia of Anthropology*. John Wiley & Sons, Ltd. DOI:10.1002/9781118924396.wbiea1758.

Cuny, F. (1983) *Disasters and Development*. Oxford, Oxford University Press.

Faas, A. J. & Barrios, R. E. (2015) Applied Anthropology of Risk, Hazards, and Disasters. *Human Organization*. 74 (4), pp. 287–295.

Firth, J. R. (1959) *Social Change in Tikopia: Re-Study of a Polynesian Community After a Generation*. New York, Palgrave Macmillan.

García-Acosta, V. (2018) Anthropology of Disasters. In: Callan, H. (ed.). *The International Encyclopedia of Anthropology*. New York, John Wiley & Sons, Ltd., pp. 1622–1629.

García Acosta, V. & Musset, A. (dir.). (2017) *Les Catastrophes et l'interdisciplinarité: dialogues, regards croisés, pratiques*. Investigations d'Anthropologie Prospective, Laboratoire d'Anthropologie Prospective, Université catholique de Louvain. Louvain, Editorial Academia-L'Harmattan.

Giordano, C. & Boscoboinik, A. (eds.). (2002) *Constructing Risk, Threat, Catastrophe: Anthropological Perspectives*. Fribourg, Fribourg University Press.

Guber, R. (coord.) (2018) *Trabajo de campo en América Latina. Experiencias antropológicas regionales en etnografía*. Buenos Aires, Sb Editorial.

Gugg, G., Dall'Ò, E. & Borriello, D. (eds.). (2019) *Disasters in Popular Cultures*. Rende, Italy, Sileno Edizioni.

Henry, D. (2005) Anthropological Contributions to the Study of Disasters. In: McEntire, D. & Blanchard, W. (eds.). *Disciplines, Disasters and Emergency Management: The Convergence and Divergence of Concepts*. Emmitsburg, MD, Federal Emergency Management Agency, pp. 111–123.

Hoffman, S. M. (2017) Disasters and Their Impact: A Fundamental Feature of Environment. In: Kopnina, H. & Shoreman-Ouimet, E. (eds.). *Routledge Handbook of Environmental Anthropology*. London, Routledge, chapter 16.

Hoffman, S. M. & Oliver-Smith, A. (eds.). (2002) *Catastrophe & Culture: The Anthropology of Disaster*. Santa Fe, Nuevo México, School of American Research.

Hsu, M., Howitt, R. & Miller, F. (2015) Procedural Vulnerability and Institutional Capacity Deficits in Post-Disaster Recovery and Reconstruction: Insights from Wutai Rukai Experiences of Typhoon Morakot. *Human Organization*. 74 (4), pp. 308–318.

Keesing, F. (1952) The Papuan Orokalva vs. Mt. Lamington: Cultural Shock and Its Aftermath. *Human Organization*. 11 (1), pp. 16–22.

Konrad, H. W. (1985) Fallout of the War of the Chacs: The Impact of Hurricanes and Implications for Prehispanic Quintana Roo Maya Processes. In: Thompson, M., Garcia, M. T. & Kense, F. J. (eds.). *Status, Structure and Stratification: Current Archaeological Reconstructions*. Calgary, The University of Calgary Archaeological Association, pp. 321–330.

Krotz, E. (2018) Anthropology in Mexico. In: Callan, H. (ed.). *The International Encyclopedia of Anthropology*. New York, John Wiley & Sons, Ltd. DOI:10.1002/9781118924396. wbiea2115.

Lavell, A. (2017) Preface. In: Marchezini, V., Wisner, B., Londe, L. R. & Saito, S. M. (eds.). *Reduction of Vulnerability to Disasters: From Knowledge to Action*. São Carlos, Rima Editora, pp. 9–14. Available from www.preventionweb.net/publications/view/56269. Accessed January 15th 2019.

Macías, J. M. (2013) Diálogos entre la antropología y la geografía en el CISINAH/CIESAS. In: García-Acosta, V. & De la Peña, G. (coords.). *Miradas concurrentes. La antropología en el diálogo interdisciplinario*. Mexico, CIESAS, pp. 193–223.

Malinowski, B. (1936) Preface. In: Firth, J. R. (ed.). *We the Tikopia: Kinship in Primitive Polynesia*. Boston, George Allen & Unwin Ltd., pp. vii–xi.

Maskrey, A. & Lavell, A. (2013) *The Future of Disaster Risk Management: A Scoping Meeting for GAR 2015*. Available from www.desenredando.org/public/2013/. Accessed December 14th 2018.

O'Keefe, P., Westgate, K. & Wisner, B. (1976) Taking the Naturalness Out of Natural Disaster. *Nature*. 260, pp. 566–567.

Oliver-Smith, A. (2017) Adaptation, Vulnerability and Resilience: Contested Concepts in the Anthropology of Climate Change. In: Kopnina, H. & Shoreman-Ouimet, E. (eds.). *Routledge Handbook of Environmental Anthropology*. London, Routledge, pp. 206–218.

Oliver-Smith, A. (2015) Hazards and Disaster Research in Contemporary Anthropology. In: Wright, J. D. (ed.). *International Encyclopedia of the Social and Behavioral Sciences*, 2nd edition. Amsterdam, Elsevier, pp. 546–553.

Oliver-Smith, A. (1996) Anthropological Research on Hazards and Disasters. *Annual Review of Anthropology*. 25, pp. 303–328.

Oliver-Smith, A. (1995) Perspectivas antropológicas en la Investigación de desastres. *Desastres & Sociedad*. 5 (3), pp. 53–74.

Oliver-Smith, A. (1986a) *The Martyred City: Death and Rebirth in the Andes*. Albuquerque, The University of New Mexico Press.

Oliver-Smith, A. (1986b) Disaster Context and Causation: An Overview of Changing Perspectives in Disaster Research. In: Oliver-Smith, A. (ed.). *Natural Disasters and Cultural Responses*. Williamsburg, Studies in Third World Societies, pp. 1–34.

Oliver-Smith, A. (1979a) The Crisis Dyad: Culture and Meaning in Anthropology and Medicine. In: Rogers, W. R. & Barnard, D. (eds.). *Nourishing the Humanistic: Essays in the Dialogue Between the Social Sciences and Medical Education*. Pittsburgh, University of Pittsburgh Press, pp. 73–93.

Oliver-Smith, A. (1979b) The Yungay Avalanche of 1970: Anthropological Perspectives on Disaster and Social Change. *Disasters*. 3 (1), pp. 95–101.

Oliver-Smith, A. (1979c) Post Disaster Consensus and Conflict in a Traditional Society: The 1970 Avalanche of Yungay, Peru. *Mass Emergencies*. 4, pp. 39–52.

Oliver-Smith, A. (1977a) Disaster Rehabilitation and Social Change in Yungay, Peru. *Human Organization*. 36 (1), pp. 491–509.

Oliver-Smith, A. (1977b) Traditional Agriculture, Central Places and Post-Disaster Urban Relocation in Peru. *American Ethnologist*. 3 (1), pp. 102–116.

Oliver-Smith, A. & Hoffman, S. M. (eds.). (1999) *The Angry Earth: Disaster in Anthropological Perspective*. New York and London, Routledge.

Ortiz, F. (1947) *El huracán: su mitología y sus símbolos*. Mexico, Fondo de Cultura Económica.

Prince, S. H. (1920) *Catastrophe and Social Change: Based Upon a Sociological Study of the Halifax Disaster*. PhD Thesis. New York, Columbia University.

Schneider, D. (1957) Typhoons on Yap. *Human Organization*. 16 (2), pp. 10–15.

Schwimmer, E. (1969) *Cultural Consequences of a Volcanic Eruption Experienced by the Mt. Lamington Orokaiva*. Salem, University of Oregon Press.

Sorokin, P. (1942) *Man and Society in Calamity*. New York, E. P. Dutton and Co., Inc.

Spillius, J. (1957) Natural Disaster and Political Crisis in a Polynesian Society. *Human Relations*. X (1), p. 327.

Taddei, R. & Hidalgo, C. (2016) Antropología Posnormal. *Cuadernos de Antropología Social*. 43, pp. 21–32.

Torry, W. I. (1979a) Anthropological Studies in Hazardous Environments: Past Trends and New Horizons. *Current Anthropology*. 20 (3), pp. 517–529.

Torry, W. I. (1979b) Anthropology and Disaster Research. *Disasters*. 3 (1), pp. 43–42.

Torry, W. I. (1978) Natural Disasters, Social Structure and Change in Traditional Societies. *Journal of Asian and African Studies*. XIII (3–4), pp. 167–183.

Vessuri, H. & Bueno, C. (2016) Institutional Re-Structuring in the Social Science World: Seeds of Change. In: Kuhn, M., Vessuri, H. & Yazawa, S. (eds.). *Beyond the Social Sciences*. Stuttgart, Verlag, vol. 3, pp. 141–167.

Visacovsky, S. E. (2017) When Time Freezes: Socio-Anthropological Research on Social Crises. *Iberoamericana-Nordic Journal of Latin American and Caribbean Studies*. 46 (1), pp. 6–16.

Wallace, A. F. (1956) *Tornado in Worcester: An Exploratory Study of Individual and Community Behavior in an Extreme Situation*. Washington, NAS-NRC Disaster Study No. 3. National Academy of Sciences-National Research Council.

Wallerstein, I. (2003) Anthropology, Sociology, and Other Dubious Disciplines. Sidney Mintz 2003 Lecture. *Current Anthropology*. 44 (4), pp. 453–465.

Wijkman, A. & Timberlake, L. (1984) *Natural Disasters: Acts of God or Acts of Man?* London, Earthscan.

Wisner, B., Gaillard, J. C. & Kelman, I. (eds.). (2015) *Disaster Risk*. London and New York, Routledge, 4 vols. Available from www.routledge.com/Disaster-Risk-1st-Edition/Wisner-Gaillard-Kelman/p/book/9780415624206. Accessed June 17th 2019.

1 Risk and uncertainty in Argentinean Social Anthropology

Ana María Murgida and Juan Carlos Radovich

Introduction

In this chapter, we present an initial survey of theoretical and conceptual approaches and research topics in studies on risk and catastrophes or disasters in Argentinean social anthropology. From this survey and its synthesis emerges a perception of an Anthropology of Risks and Disasters, which is still under development.

In this survey, we include anthropological papers from two books published in 2015, which focused on risk and disasters in either their research subject or methodology. One of these books, *Contributions from geography and social sciences to Argentine case studies*, edited by Claudia Natenzon and Diego Ríos, is a collection of studies by researchers from the "Programa de Investigaciones en Recursos Naturales y Ambiente" (Research Program on natural resources and environment, PIRNA) at the Universidad de Buenos Aires (Buenos Aires University, UBA). The second, the book *Riesgos al sur Diversidad de riesgos de desastres en Argentina* (*Risks in the south. Diversity of disaster risks in Argentina*), edited by Jesica Viand and Fernando Briones, was the product of an open call by the Red de Estudios Sociales en Prevención de Desastres en América Latina (Social Research Network for Disaster Prevention in Latin America, LA RED). Our research was performed with a survey of the literature referenced in these two books, and with papers presented at Argentinean anthropological conferences on these topics between 2011 and 2018. Our analysis identified main research topics, theories and methods applied, the most frequently cited authors, and the disciplinary or interdisciplinary character of the research. This survey helped us to discover the emergence of a national social anthropology and revealed how the study of risk in social sciences was carried out by Argentine anthropology.

Different regions of Argentina and their urban and rural populations are periodically and significantly affected by floods, droughts, earthquakes, volcanic eruptions with their ash depositions, or by industrial accidents and pollution. Despite the recurrent nature of these phenomena, institutions are affected by unpredictability and uncertainty when it comes to applying public policies designed to prevent, reduce, or mitigate the damage caused by such events. In general, only post-disaster assistance is provided without attempts to reduce future vulnerabilities or risk.

We also need to consider risk as the result of a combination of natural and technological factors, social organization, adaptive strategies, and models of development, which altogether generate a state of uncertainty. Argentinean anthropology of risk examines this combination of factors in studies of disasters, conflicts, and crises. A disaster occurs when natural or technological hazards, socially constructed risks, and increased vulnerability of social groups gather in time and space (García-Acosta, 2005, 2006).

The anthropological approach to extreme or critical socioenvironmental situations has assumed some principles which have been constructed over several decades of research by different social sciences in a dialog with natural and engineering sciences, while sharing distinctive viewpoints with social sciences and humanities.

It is important to state that risks constitute normal components of environments known by different cultures and societies, under diverse expressions, and are faced in various manners in terms of response. Disasters commonly combine potentially destructive natural or technological agents to which a vulnerable population is exposed. The combination of natural and social elements produces various damages and losses in the principal living, organizational, and cultural elements of the affected populations. The concept of vulnerability presents difficulties in its interpretation since those sectors affected by catastrophic processes are commonly populations that have been made vulnerable due to conditions of structural social inequality. Risks and disasters have greater impact when vulnerabilities of both the environment and the communities affected come together. In this sense, the vulnerability to disasters and catastrophic processes becomes understandable in terms of the conditions of inequality and subordination that exist in certain societies.

Political and production decisions that drive development weigh relative costs and benefits, and reflect on different levels of risk, economic, ecological, technological, etc.; however, social and cultural aspects are commonly relegated to second place and are perceived as "externalities".

Perception and the level acceptance of risk are based on historical processes collectively constructed under social principles and values that determine which real dangers are recognized and faced at different times. Past social experiences of disasters are spread along individual or collective memory processes and contribute towards an understanding and realization of significance that show great variation and some discernable patterns (García-Acosta, 2004; Díaz-Crovetto, 2015).

Anthropology of Risks and Disasters incorporates theoretical approaches such as cultural and political ecology, and other subdisciplines such as economic and political anthropology. It also builds on applied studies of refugees or relocations under different models of development (dam constructions, oil and gas explorations, expansion of the agricultural frontier, etc.), environmental impact studies, and work on climate change and the management of risk (Oliver-Smith, 2002; García-Acosta, 2006). Studies on risk and disaster, whether on its theoretical construction or on its applications, are the result of interdisciplinary dialog and of needs for management and design of public policy (Oliver-Smith, 2002; Warner et al., 2010; Baez Ullberg, 2017a; Murgida, 2012).

The 1980s was an important period for the development of the discipline, when a theoretical focus on management facilitated the study of disasters as a complex phenomenon in multiple contexts, beyond the purely physical models with their tendency to inductive analysis (Bartolomé, 1985; Ribeiro, 1985; Hansen & Oliver-Smith, 1982; García-Acosta & Suárez, 1996). This new perspective implies that sociocultural, economic, and political conditions prior to an event affect society before, during, and after a catastrophe. As a result, more operational concepts of vulnerability, exposure, adaptive strategies, and risk were introduced to analyze affected populations (Radovich, 2013; Valiente & Radovich, 2016).

Among the first social disciplines in Argentina that included risk in their analysis mention may be made of geography, sociology, and later anthropology. The first lines of research followed authors from the school of political economics of disasters, from the Disaster Research Unit of the University of Bradford, and of the Disaster Research Centre from the University of Delaware under the leadership of the sociologist Enrico Quarantelli with his analysis of social responses that applied "symbolic interactionism" to field studies. Academics from geography, sociology, and social anthropology converged in 1992 to form a network, LA RED, within the framework of the UN efforts on disaster reduction. This contributed to relevant research and works for the development of a focus on disaster risk in Latin America.

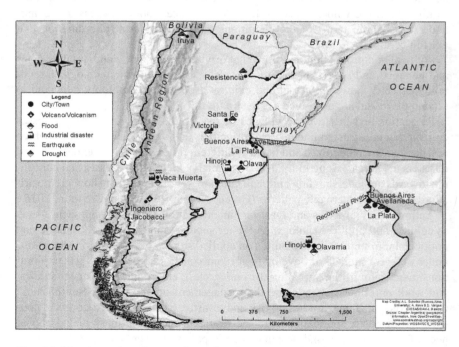

Map 1.1 Map Argentina. Case studies and main areas mentioned

We have found case studies on disasters and examples of conflicts related to the perception of risk. Among these, a reduced number of studies focused specifically on risk. Other examples to mention approached the situation of risk or disaster in an implicit or emergent way from a problem-centered research focus. Yet, the material collected and analyzed allowed us to establish a framework to assess the state-of-the-art in Argentinean Anthropology of Disasters.

Background to risk and its inclusion in research

The initial object of study of anthropology, which then was called ethnology or cultural anthropology, were the ethnography of societies that were conceptualized by Marcelo Bórmida (1970) as "primitive cultures", or as surviving relics with their central categories of community, tribe, and culture. It needed historical global transformations such as the decolonization process in Africa and the critical reformulation of the initial paradigms of British Social Anthropology or French ethnology and sociology, like that of Maurice Godelier and Georges Balandier, for anthropology to recognize the transformative effects on both its objects of study and the way it was approached and conceptualized. Since then, anthropologists have referred to it as Social Anthropology, which means, addresses the "social" and on "society" issues.

Around the end of the 1950s, Argentinean Anthropology was still engaged along the lines of archaeological investigations and ethnology of historic and cultural origin, and the physical anthropology and folklore from Hispanic tradition. These lines allowed inventories of cultural groups believed to be on the brink of extinction. However, due to the influence of scientific sociology, which required that economic, political, social and cultural aspects be included, together with anthropologists who studied abroad, led to the development of social anthropology. Until then, this subject matter was taught at the Department of Sociology at the UBA, where it included cultural diversity as part of an "us", as part of the contemporary Argentine society (Guber, 2018; Balbi, 2012). Since the middle of the 1960s, Argentinean anthropologists began to study the living conditions of different cultures, native peoples, *criollos*, small producers and the urban marginal populations in environments modified by development politics of the time. In this way, a group of professionals developed a theoretical and political commitment towards resolving real, concrete problems as noted by Hugo Ratier (1971), one of the pioneers of Social Anthropology in Argentina. These researchers considered risk as part of their questioning; however, this topic was not a central axis of the analysis then.

Ratier recounts how, in those years, anthropologists, psychologists, social workers, educators, architects, and physicians applied their disciplines to an extension project which UBA conducted on the Isla Maciel, a community in the oldest industrial suburb of Buenos Aires, Avellaneda. Internal rural migrants and those from different countries arrived in this community on the margins of a malodorous creek (*Riachuelo*) full of the effluents from shipyards, slaughterhouses, tanneries, and port activities, among others. Newly arrived vulnerable families settled in this polluted urban environment with health risks, and these migrants tried to make the best of the adaptive capacities that resulted from the social and productive relationships of their regions of origin (Ratier, 1971; Bartolomé, 1985).

In the Universidad Nacional de Misiones (National University of Misiones, UNaM), in the northeast part of Argentina, the anthropologists Leopoldo Bartolomé and Carlos Herrán headed working groups that evaluated the social and cultural aspects of forced displacement resulting from the construction of the binational (with Paraguay) hydroelectric dam of Yacyretá. This work was done in the mid-1970s during the last military dictatorship (1976–1983) in the context of one of the World Bank projects in the region. Their work contributed to the planning of resettlements, and, through improving the communications on the project, managed to reduce the uncertainties among the population in the face of a development project that would change their lives and social organization. The working groups' mode of operations initiated an applied anthropology by establishing a dialog between anthropologists, engineers, government representatives, and the affected population. Findings contributed to increased knowledge on the sociocultural and economic conditions when selecting resettlement sites.

These researches were conducted in the setting of a global network of researchers that worked on similar problems linked to large development projects and resettlements in a complex society. These included for example the anthropologists Ángel Palerm, Miguel Bartolomé, and Alicia Barabas in México; Eric Wolf, Thayer Scudder, Elizabeth Colson, William Partridge, and Michael Cernea in the United States; and Silvio Coelho dos Santos, Cecilia Vieira Helm, Maria Jose Reis, and Gustavo L. Ribeiro among others in Brazil. Amongst the Argentinians who had studied at UBA and returned to their country after postgraduate work abroad, were Esther Hermitte with a PhD from the University of Chicago, Eduardo P. Archetti from L' École Pratique des Hautes Études, Paris, Hebe Vessuri from the University of Oxford, Leopoldo Bartolomé who had studied in the University of Madison, as well as Santiago Bilbao and Hugo Ratier who studied at UBA, but had to go into exile during the military dictatorship. The careers of these anthropologists were important for linking the development of international science to the real problems of Argentina. This permitted a combination of different economic and political critical perspectives to contribute to an understanding of how the distribution of power, inequalities and vulnerabilities were constructed, and what role they played in connecting different actors in society (ethnicities and classes with their modes of production).

For some of these authors, the notion of 'social articulation' was the guiding principle in the analysis of social relationships with their adaptive structures for risk in the context of uncertainty. In this way, the experiences of development are reinterpreted as perceptions of hazards and risks. Hermitte and Bartolomé (1977) proposed that the rural peripheries of capitalism are subsidiary to the cities, arguing that from a systemic viewpoint both are tied together through social linkages (Mastrángelo, 2000). The uncertainties associated with the precarious ownership of land, the resettlements, and the effects of territorial intervention reduced the viability of traditional adaptation strategies of the affected population. Studies on producers and indigenous populations (Briones & Olivera, 1989; Radovich & Balazote, 1996, 1998), an on marginalized urban groups (Ratier, 1971; Bartolomé, 1985), showed them as defenseless against the seasonal floods or those produced by land-use change and the impact of a large-scale dam (Catullo, 1986; Catullo & Brites, 2014).

During that period, geographers and sociologists conducted ex-post studies on urban flooding in different regions. They analyzed the socioeconomic structures and demography in the context of physical-natural causes identified by scientists from diverse disciplines and taking into consideration the government responses to such events occurred. Those initial studies showed that political institutions normally considered such disasters as extraordinary events without recognizing their roots in ongoing processes and recurrent phenomena. Consequently, policies did not include this research or any analysis of risk management into land-use planning, nor did they recognize that the degree of vulnerability of the populations determined the dimensions of the disaster.

Later on, some studies have shown that scientific and political uncertainties about vulnerability and disasters, increase risk and reduce institutional capacity of response. Because of these studies, analysis of risk management and emergencies gained greater importance, while a public debate on the need for preventive measures by governments began to develop. This was accompanied by the recognition that these were not isolated events, but critical moments of a larger process of risk, which includes decisions on development (Herzer, 1985; Natenzon, 1995, 2003; Gentile, 1993; González, 1997; Merlinsky, 2013, 2016, 2017).

Questioning the exceptional nature of disasters was already part of the anthropological analysis by Bartolomé. His studies showed that "floods, despite their frequent occurrence, had been treated as dramatic events, in a way unique [exceptional]'emergencies' which end in themselves and require [only improvised] exceptional measures"(Bartolomé, 1985: pp. 7–8). Argentinian anthropologist Juan Carlos Radovich, analyzing the political-institutional responsibility at both national and international levels, questions the role of some international agencies which on one side support territorial interventions which vulnerabilize the population and at the same time recommend remediation frameworks (Radovich, 2013).

As an example, we can mention early initiatives of the World Bank, which in 1980 included the issue of social vulnerability (as an undesired effect) in development projects, and provided resolutions that referred specifically to indigenous populations affected by large-scale projects. Similarly, in 1987, United Nations recognized the importance of social impacts of the predominant development model and the centrality of the problem of risk, which influenced subsequent declarations and conventions. This situation was formalized in Argentina through the signing the Hyogo Declaration (2005) and its action framework, the Sendai framework for the reduction of risk and disasters (2015). Coordination and cooperation between institutions engaged on risk and disaster managements was formally institutionalized in 2004 with the creation of the directorate for civil protection (later to become a Secretariat in 2014). This evolved further with the

Integrated Program for Emergencies and Crises, and the most important milestone is the creation of a National System for the Integrated Management of Risk & Civil Protection by Law N° 27.287 enacted in 2016. Because of space limitations, we only briefly mention the institutional role of the Ministry for Science, Technology, and Productive Innovation (demoted to Secretariat in 2018), which was fundamental in leading since 2012 the process of scientific interdisciplinary and interinstitutional dialog to organize the participation of different actors in managing useful information for emergencies and catastrophes.

The emerging linkage between science and decision-making is rooted in the early advances in the field of social risk and disasters, which contributed, to the academic and professional learning of young researchers at Gino Germani Research Institute, Faculty of social sciences from the UBA, and to the broader dissemination of outcomes on the continent. In 1992, the sociologist Hilda Herzer joined the interdisciplinary group of LA RED from where she also contributed to the development of national inventories of disasters. This initiative brought together, and made available in Spanish many of the advances and research works from Latin American countries, providing an important point of reference to those who were starting research in Argentina. Another important place of learning and professional practice is the aforementioned program PIRNA at the Instituto de Geografía (Institute of Geography) in the Facultad de Filosofía y Literatura (Faculty of Philosophy and Literature) from UBA, led by Natenzon, which developed applied investigations in a dialog with natural, exact sciences, and with management organizations. It should further be noted that institutes from the Universidad Nacional de Cuyo (National University of Cuyo) and the Laboratory of Desertification and Land Reform of the Instituto Argentino de Investigaciones de las Zonas Áridas (Argentinean Institute for Research of Arid Zones) in Mendoza played a significant role. These spearheaded interdisciplinary research with a strong physical component in a region that is exposed to a variety of hydroclimatic and seismic risks. The work of the geographer Elena Abraham (2009) provides a key reference on the topic of environment and desertification in Latin America and the Caribbean.

In all these teams, scientists worked with a perspective of vulnerability of the society and its institutions with reference to the ideas by Allan Lavell (1996, 2005) and Andrew Maskrey (1993). The development and management of urban lands exposed to risks and disasters, particularly floods, was studied by Caputo et al. (1985), Herzer (1990, 1993), Herzer & Federovisky (1989), Gentile (1993), Natenzon (1995, 2003), Balazote (1997, 2001), Sarlingo (1995), and Merlinsky (2013). Since the late 1980s, UBA implemented the first postgraduate courses from a perspective of the social sciences on the available knowledge on, and management of natural disasters, directed not only at academics but also towards contributing to governmental institutions. Ever since, the voice of social sciences was part of governmental interinstitutional debates on risk management, disasters, catastrophic processes, and eventually on climate change.

In 1995, Natenzon, as part of the aforementioned initiatives, brought together the ideas of Silvio Funtowicz, Jerome Ravetz, and Ulrich Beck towards a model that permitted the analysis of different dimensions of risk, potential hazard, vulnerability, and exposition, which all are conditioned by uncertainty (scientific and political) (Natenzon, 2015). The perspective of post-normal science of Funtowicz and Ravetz (1993) not only proposed an interdisciplinary dialog but also included a participative methodology for scientists and decision-makers at different levels.

The growth of such applied science saw an interesting contribution by Héctor Poggiese who was trained in project management and who together with the sociologist María del Carmen Francioni developed models of "associated management". These were based on the use of methodological tools to organize and monitor the development of processes of planning and management by government or private entities in a participative way that involved all actors. In this process, risk was one of the factors to consider and uncertainty is reduced by the inclusion of different types of knowledge (Poggiese & Francioni, 1993).

Much of the works produced by research teams within the framework of applied science and consulting services is found in reports or in the "gray literature". Nevertheless, professional practice served as input for academic analysis and deeper reflections.

Recent anthropological research

In our review of recent scientific literature, we find that most of them are case studies presented from their specific disciplinary framework in which the common denominator accounts for both an anthropological and ethnographic approach. This refers to a construction of knowledge, which integrates native perspectives and those of the researchers, and therefore recovers these significant contexts of reference that are shared between social actors and agents (Guber, 2018; Visacovsky, 2017).

When dealing with social risk, catastrophes and critical situations, anthropologists define them as events or process, which in a way reflects the conceptual proposal by Susanna Hoffman and Anthony Oliver-Smith when they define a disaster as

> a process/event combining a potentially destructive agent/force from the natural modified or built environment and a population in a socially and economically produced condition of vulnerability, resulting in a perceived disruption of the customary relative satisfactions of individual and social needs for physical survival, social order, and meaning.
>
> (Hoffman & Oliver-Smith, 2002: p. 4)

Events and processes are part of the same phenomenon; the publications may place more emphasis on one or the other, but both are always present. In the first case, the focus is centered on the phenomenon development process in the short term. However, in most of the cases, the processes analysis focus takes the historical genesis of the phenomenon in the long term, emphasizing in a relational way the construction of risk and social vulnerability. Their analysis analyzes political and juridical devices that demonstrate the construction of risk and the weaknesses of the institutional system to articulate and implement preventive or responsive actions aimed at reducing risk and mitigate sudden or slow disasters.

Social perceptions are analyzed as a function of different levels of local or academic knowledge, together with the actions and effects produced, as much in the construction of risk as in the effectiveness of adaptive strategies in critical situations.

Among the central approaches taken by different authors, there is the historical process analysis, which establishes a dialog between past, present, and future. It is significant that most of the publications show the origins of material transformations, which in most cases are linked to development policies. This approach also shows changes in the perceptions of actors involved, and in the relation of different social groups with the future and the way it is constructed.

The problematizations or ways of formulating observed issues in terms of scientifically grounded problems are developed in accordance with theoretical/methodological trends in social sciences and their linkages to the context in which they were produced in different projects, be they disciplinary or inter- or trans-disciplinary. In more general terms, following Virginia García-Acosta (2018), we find that the trends underlying anthropological analysis are the critical theories such as the political ecology or Marxism, the theories of systems and complex systems, and the structural constructivism.

Generally, the storylines on risk, catastrophes, and vulnerabilities are inspired by critical theory or political ecology, with central axes of research such as these: social vulnerability, inequality, poverty, material and symbolic memory, bureaucratic procedures, acceptability of risk, conflicts, physical-social amplification of hazards, threats or dangers, and the forms in which they are perceived, incorporated, or denied in practice and in land use and management.

The systemic approaches within the framework of critical theory or when anthropology interacts with other disciplines, account for sociocultural, material and symbolical complexity of the entire phenomenon under study. In these works, changes are considered as inherent to systems and imply further changes in one or several main aspects of social and cultural organization. At times, change is also expressed in conflict because of contradictions between needs and interests.

Such contradictions always have territorial or institutional roots where vulnerabilities, in which hazards and uncertainties that define risks of catastrophe or crisis, converge simultaneously. At the same time, such contradictions manifest themselves in the way that they leave different social actors in a vulnerable position, particularly those in poverty, while favoring the wealthier sectors of society.

Argentinian anthropologist Ana Murgida (2012) explored social and environmental changes along the agricultural frontier of the Chaco in the Argentine Province of Salta under climate change, as well as modifications in patterns of land use for agricultural production. She examined how these changes brought opportunities for present and future productivity. The author examines the links between the social relationships of production and the hydroclimatic changes, and how these links were being perceived. She identified political practices related to the distribution and appropriation of land and the social vulnerability resulting from processes that changed natural ecosystems and environmental conditions. These changes were the results of political and economic decisions that determined access to productive land and the conditions of production, and affected the social distribution of available resources. Amongst the principal consequences, the authors identified forced displacement of farmers and aborigines towards the margins of rural villages and towns, and the accumulation of lands and profits amongst agricultural entrepreneurs (Murgida, 2012, 2013; Murgida et al., 2014).

Technical innovation is a conceptual resource available to those sectors with access to capital, and it determines the distribution of means for improving profitability. This link between capital and innovation accentuates the differential access to resources and increases iniquity and vulnerability. There is a clear mismatch between the availability of state-of-the-art technology and the ability to make use of it for small and medium producers with limited financial, social, and cultural resources. This research showed that political activity does not consider social issues in the monitoring of norms and implementation of development projects and technologies such as the construction of canals (Murgida, 2012; Murgida & Gentile, 2015; Murgida et al., 2017) and the implementation of new irrigation systems (Riera & Pereira, 2015; Riera, 2017).

When the foundations of political decision-making conflict with scientific evidence and with the perceptions of those social actors who are exposed to risk, uncertainties permeate development activities, and conflicts arise which may need to be resolved on a judicial level. This was shown for a case of potential electromagnetic contamination and its impact on health by the construction of a high voltage powerline in the town of Ezeiza (Province of Buenos Aires). The construction plans precipitated the organization of the social protest and the judicialization of the resulting conflict. It resulted in the application of the precautionary principle in the verdict on an environmental contamination (Murgida, 2000).

The social organization that responds to the perception of risk promotes cognitive frameworks in which information is produced and communicated and in which productive practices are appraised. This was observed in the case of Hinojo, a small town located near Olavarría city, in Buenos Aires Province, in which the population exposed to the dust generated by a fertilizer plant became organized and began questioning the standards applied by the government for the chemical and physical composition of air that affects human health. Agustina Girado and Rosario Soledad Iturralde, both social anthropologists (Girado & Iturralde, 2015), as well as Iturralde (2015), contextualize the situation in the historical transformation of the landscape. The authors analyze how, in a situation of risk, historical-cultural factors are re-signified, through which the productive processes of the region are perceived, once environmental conditions begin to threaten public health and the quality of life of the population. In this way, they exposed the contradictions between the local characterization of the environmental conditions as a disaster and the governmental lethargy, which was perceived as indifference for not attending the complaints of residents faced with a situation of uncertainty in their conflict.

However, when governmental institutions recognize situations of exposure to danger and disasters, their answers vary as a function of the vulnerability of the affected populations. This is clearly shown in the work by Débora Swistun, anthropologist, and Javier Auyero, sociologist (Auyero & Swistun, 2007; Swistun, 2015), who addressed the historical construction of a disaster in slow motion caused by air, water and soil pollution. The authors directed their efforts at the vulnerability of a poor urban population that lives in an area of petrochemical industries in Dock Sud, in the municipality of Avellaneda (Buenos Aires Province) close to Buenos Aires city. In analyzing life histories, they address implicit and explicit uncertainties in the present and historical perceptions of the inhabitants and of health professionals concerning pollution risk factors and how they affect people. In this way, they show how uncertainty is core to the construction of scientific and political knowledge that creates a "condition of waiting" without terms for adequate answers, even though all involved actively participate in an interinstitutional dialog.

At times the social perception of risk and vulnerability as a result of processes of contamination find no echo in public agendas, or, if they do, it is because of international agreements which are in some way implemented even if only formally. This is the case in the construction of a public agenda on the problem of air quality in the city of Buenos Aires (Murgida et al., 2013; Abrutzky et al., 2014). The authors analyzed the phenomenon using available official data on morbidity and mortality and observing the incipient instances of intersectoral participation in evaluating the problem and acting on mitigation. Amongst the results stood out the importance of international agreements, which demanded formal action, but also the difficulty to implement integrated policies which included social problems being limited to the implementation of improved monitoring of air quality.

An ethnographic approach allows us to recover the collective memories of social constructions and the distribution of vulnerabilities, hazards, and risks, revealing institutional bureaucratic mechanisms that accompany such processes. In this way, Baez Ullberg, a Swedish anthropologist with very important contributions to the study of Anthropology of Disasters, addresses the material memory of disasters that turn into a social and political enquiry which, amongst other issues, reveals the limitations of diagnostics and public works that result from discontinuities in staff members, funding, and focus (Baez Ullberg, 2015, 2017b). The author shows these findings based on a post-disaster analysis of the catastrophic flooding of 2003 in the city of Santa Fe, in the homonymous Province. She probes the construction of material memory, or memory scape, of disasters, and how this construction operates in different social groups and institutions involved in a catastrophic process. In this way, it reveals bureaucratic mechanisms of concealment and selective memory which are activated not only upon recurrence often inundation, but also with any change of government, thus debilitating the effectiveness of policies for management and the reduction of risk disaster.

Diego Zenobi (2015, 2017) analyzed the social, political, and judicial dimensions of a catastrophe known as the "Tragedia de Cromañón" ("Tragedy of Cromañón"), the fire that destroyed a discotheque in Buenos Aires city in 2004. The author analyzes the different meanings and practices that the memory took on among the grieving government officials and judicial representatives. Even though there was an explicit consensus on the causes of the disaster and the administrative mechanisms that conspired against preventing it, the judiciary increased the tragedy for the grieving and survivors by submitting them to drawn-out probes into their losses and sufferings. In this work, the author highlights tensions between the actual bureaucratic and institutional treatment of suffering, addressing that social vulnerability is incorporated in a politicization process of the memory of the victims and their relatives.

In this research, one can observe how and when the politicization of the bureaucracy interacts at different levels of government and limits the continuity in public administration, especially in the prevention and management of risk. The consolidation of behaviors of forgetting or of concealing information basically occurs at the point of transition between governments of different political orientation, reflecting in a way the weakness of the institutions in the country.

Historical relational analysis, which applies tools of discourse analysis and social network description, provides opportunities to compare results and scientific proof within interdisciplinary work; thereby facilitating dialog and collaboratively validating results obtained with different methods. In this way, one can incorporate social significance into data and results from natural sciences and at the same time examine the relations between social actors in the sciences and those involved in the management and construction of knowledge. This provides for a significant linkage between results and explanations from the social sciences and other disciplines (such as atmospheric sciences or hydrology) integrating the analysis of relations with those between actors and the environment. Ethnographic analysis can reveal the perception of exposure and social vulnerability as well as the traditional adaptive strategies applied by the local population to face risks and prepare for disasters (Murgida & Kazimierski, 2017; Murgida, 2012).

Local knowledge concerning behavior of mountain rivers appears clearly in the research of Murgida and Gasparotto (2015) on the village of Iruya in Salta Province. Among the characteristics highlighted by local population and administrators were the seasonal increases of flow, which erode the riverbanks which affect villages. Areas of major exposure to cresting waters were identified and mapped together with the description of social vulnerability and the ancestral strategies of mitigation that had fallen into disuse, and thereby opened the way for external proposals from government organizations. As a result of data systematization, a dialog channel was provided between different social actors with the aim to reduce uncertainties and advance towards the design of defensive structures and early warning systems.

Uncertainty linked to the management of flood-prone lands is presented by Lucila Moreno (2015). The author examined the problems that exist between management teams and the affected population in arriving at agreements on land use in urban flood zones, and the possible displacement of inhabitants of precarious settlements in the basin of the Reconquista River (Province of Buenos Aires). She analyzed how contradictions deepen between hazards and risks identified in the context of flooding, and the vulnerability perceived by the affected population before an eventual future relocation given their precarious land tenure and a potential lack of political leadership. An additional result of this study showed how often, in an interdisciplinary context, the role of the anthropologist is also one of mediator between risk perceptions by inhabitants and government administrators.

Floods have a major presence among the cases considered here. They represent a significant problem in management and a social problem carved into social memory, expressed in the shaping of landscapes and in native narratives, and refreshed in the collective memory by the recurrence of a phenomenon (Baez Ullberg, 2013). At present they address an important place in political agendas because they are problematized by different sectors of society from those directly affected to the politicians and scientists who produce information since it is now recognized as common sense that disasters are not natural. This is clear in at least two catastrophic examples in two important cities in Argentina, Santa Fe (occurred in 2003) and La Plata (occurred in 2013), which felt the impact of flooding both in poorer and well-off neighborhoods.

In both cases, three factors converged in the catastrophe, a large population, exceptional rainfall, an increase in surface water level, and a lack of appropriate infrastructure to mitigate or prevent the flooding. In both cities, thousands were affected, and dozens died. Social conflict which accompanied the post-disaster phase remained active through legal actions, almost at the end of the second decade of 21stcentury, which showed results in terms of addressing legal liabilities (Etchichury et al., 2016; Baez Ullberg, 2015, 2017b).

The research reveals, for these and other cities, those patterns of floods share common physical and infrastructural causes. Nevertheless, investigations from social sciences highlight the importance of historically accumulated social vulnerability as a decisive factor shaping a catastrophe, and conditions of delays to which the poorer segments of the population are exposed in such cases.

When we add an historical approach for a reading of the selected case studies, risk management evolution becomes visible. In Argentina during the 1970s, we can seea policy of post-disaster help, with the absence of plans for the prevention of future catastrophes, while there is an exclusive focus on a response during the event. Sarlingo (1995), Balazote (1997, 2001), Roze (1997), and Boivin et al. (2000) analyzed catastrophic floods in Olavarría (Buenos Aires Province), Chaco city (Chaco Province), and Victoria Department (Entre Ríos Province), respectively. They paid attention to poor, affected populations, who received attention from the state during the last military dictatorship. In the city of Resistencia, the capital city from Chaco Province, the historical process of development resulted in the settlement of the poor inhabitants in areas of floods exposure. When disasters occurred, local and provincial governments responded in a way that was both regulatory and stigmatizing for the poor sectors affected, reflecting the dominant ideology and institutional practices of the time. In the military headquarters, rights to self-determination by the poor were abrogated. Without voice or rights, they were essentially infantilized, receiving only the most basic necessities, while those who stayed outside the controlled areas were treated as delinquents. The author states that this management of catastrophe does not allow any improvement in the situation and with every repeat during subsequent flood events, the vulnerability of the poor populations is increased as they are exposed to the loss of their belongings, while authoritarian mechanisms prevent them from exercising their citizens' rights. Neither risk nor adaptation appear in the state responses, or in the analyses of the process. Nevertheless, formal advances in risk management and disaster response, floods, and the lack of required measures continue to have repercussions in the lives of the inhabitants.

An anthropological study carried on in La Boca neighborhood in Buenos Aires city in 1994 showed recurrent floods and gave new insights (Suárez, 1994). This study revealed additional problems that the local population experienced during floods: environmental pollution, fires in precarious housing with inadequate electricity connections, etc.

The author demonstrated the efficacy of the strategies of adaptation and response during the emergencies by the affected population, such as the existence of autonomous social networks, which disseminated the necessary information in an early warnings about the southeast winds which cause floods in the area. This allowed them to make better use of government action still less developed for the management of risk such as evacuation and the food provision.

A decade later, Murgida & González (2004), continuing with the same case of floods in La Boca district, combined methodologies from geography and anthropology to map exposure and social vulnerability through indicators from census data. In this way they analyzed the social and cultural construction of vulnerability as well as strategies of adaptation and mitigation undertaken by the neighbors themselves and by governmental institutions. Anthropological work covered different actors' perceptions of flood impact and constructions for flood defense to mitigate the effects. The authors concluded that the institutional vulnerability is associated with political discontinuities resulting from the political cycle of changes in administration, which interrupted management processes and public works. When analyzing design and planning activities, it became evident that there are both community and government values regarding the engineering solutions and calculations to define mitigation measures, and also a weak dialog with other disciplines from natural or social sciences which addressed the phenomenon of floods in the context of climate change and disaster risk reduction. In this way, the authors also showed that neglecting the problem of uncertainties related to future security, which imposed an additional risk factor because it generated a false sensation of security associated with public works, although other scientists of different disciplines provided information on physical and social characteristics of flood phenomena concerning to climate change and local knowledge (Ríos & Murgida, 2004; Murgida & Natenzon, 2009).

Argentinian anthropology incorporates studies on volcanic ashes in the 21st century, particularly regarding catastrophe caused by the dispersion of volcanic ash in 2011 in the region of Patagonia. This is reflected in the studies carried out by different disciplines including natural and social sciences. Archaeology provided a previous study by Luis Borrero (2001), who looked at the impact of volcanoes in historical and ethnohistorical terms, considering the general uncertainty that surrounds both anticipation and consequences of eruptions in the Andes and the area of the Argentinian Patagonia. Researchers who analyzed the risk stemmed from this phenomenon later recovered his contributions.

One of these studies was carried on by Radovich (2013), who analyzed the problem as a social drama which revealed the lack of foresight in public policy as much as the autonomous responses of indigenous Mapuche communities who live in the Patagonian region. The author addresses the value that different social groups give to adaptive strategies to confront the effects of volcanism, concentrating on the revaluation of religious ceremonies by Mapuche communities. He showed how cultural efficiency can be attributed to ancestral knowledge on how to maintain production in degraded ecological systems, as vulnerabilities due to enclave-type activities in Norpatagonia (Valiente & Radovich, 2016). However, he also considered the origin of local phenomena that face situations of risk or disaster, showing up institutional storylines at multiple levels of jurisdiction, administration, sectors, and socioeconomic factors, which all combine to define the catastrophic process.

The historical evolution of vulnerability at the Patagonian steppe is an ongoing disaster in that region, and has been since the founding of the nation. It is thoroughly analyzed by Murgida & Gentile (2015) through the transformations of the vulnerability and the acceptability of risk by Mapuche and *criollo* producers under the periodic droughts. The authors began their fieldwork during a prolonged drought (approximately seven years), and a time of a major volcanic eruption that deposited large amounts of ash on the land. In this context, they analyzed adaptive strategies from the community and how they adopted those strategies by different state levels. The authors carried out a comparative study between different types of policies applied in the region, and the results of these policies. Among their findings, they found that political and juridical measures applied between 2006 and 2014 improved the strategies of animal production and marketing by strengthening the networks of social solidarity between Mapuche and *criollo* communities and improved the dialog with the government. This resulted in a stronger exercise of their rights, and, consequently, modified the thresholds of acceptability of risk associated with prolonged drought, snowfall, and volcanic ashes exposure.

The aforementioned studies show the importance of anthropological analysis in linking physical environmental components with the interactions between risk perception, social vulnerability, political development, and institutional culture in order to understand the forms in which risk is faced.

The analyses were further developed in a transdisciplinary approach between anthropologists, geographers, and government functionaries, who contrasted census indicators of vulnerability with an analysis of public policy aimed at addressing the material and cultural needs of communities. This was done before a scenario by which between 2005 and 2015 public management moved from direct assistance during a disaster to the promotion of more long-term measures, including the provision of access to public services, the regularization of land tenure, and the design of development programs.

In this way, the last disaster at the steppe (2006–2014) initiated a slow process of including vulnerability in the management of the territory and development of the Patagonian steppe through participative programs of community-based design.

The dialog between academia and public institutions presented in this study confirmed the needs to include risk management in policies aimed at rural and indigenous populations (Murgida et al., 2016, 2017). With a change of national government at the end of 2015, this process suffered a setback, both in the matter of rights and in the interest to reduce social vulnerability. This confirmed the implementation of concealment mechanisms deepening and ideology which naturalizes vulnerability and is linked to changes in public administration,

Poverty, precarious land tenure, and a fragile legal foundation for exercising rights are main constituents that characterize social vulnerability in situations of disaster risk. They also reveal the character of state interventions, as they operate to reduce or increase vulnerability particularly in catastrophic situations.

Final reflections

From this review, one can see that, during the last decades, Argentinian anthropology has addressed the problem of risks and disasters in several different ways. In the majority of the studies done in Argentina and linked to risk and disasters, we found heterogeneous studies around critical events that both affected all aspects of human and social life and disrupted order and normality. Each author carried out research from different ethnographical approaches, but where its own disciplinary identity remains. Nevertheless, the examples which we have collected in this chapter show different facets of social and anthropological research on risk and disasters which, in combination, constitute an "Anthropology of Risk" in an environmental, social, and political context specific to Argentina.

Work on research and the reflections shared in environments of intersectorial dialog on risk and disasters allows us to affirm that anthropologists have made a relevant contribution through placing research in the field of knowledge in terms of scientific work, management and professional consulting.

Using ethnographic methods and techniques like participant observation and the development of theory linked to a holistic and comparative focus, anthropology has provided information on the encounter with protagonists of the topic approached. Direct encounter with agents and its record provide elements that strain with the view of the researcher for analyzing phenomena, accounting for the intricacy of political, organizational, structural, technological, and cultural and identity dimensions, and define the distribution of risk and the effects of disasters.

The questions that were asked and their analysis account for conceptual linkages between risk and disaster by providing a historical construction or sociocultural origin in the territory of institutions, knowledge, and uncertainty. The works carried out in different contexts and with approaches that deal with processes of long or short duration manage to account that social and cultural values of the actors involved – with different degrees of power and vulnerability – affect the decision-making in general, and of government institutions particularly.

In these studies, one prevalent concept is that of uncertainty. This refers to imprecision at the frontier between knowledge and certainty, and shows up in different fields as a lack of information which makes affected communities and their institutions vulnerable at the time of integrated responses. In this way, uncertainty can be defined as a gap in the understanding of risk, disasters, and their effects, among agents in the territory, which limits their capacity to make appropriate decisions.

Anthropologists include in their studies the tensions between knowledge repertoires and action logics linked to situations of risk or disasters, which account for the treatment of uncertainty when making political-institutional decisions. The tensions between the ways of perceiving situations of risk and vulnerability reflect a "rupture of intelligibility", which places at the center of the analysis the encounter between uncertainty regarding adaptive strategies and cognitive function in critical situations, both unexpected and recurrent. With the analysis of historical transformations, uncertainty is shown as a form of imperfection in the information that is considered reliable. It impacts on the effectiveness of the adaptive structures of the communities, as well as on the institutions involved in the management of risks, disasters, and conflict resolution.

As previously affirmed, the Argentinian anthropological studies mentioned were allowed to distinguish different degrees and types of uncertainty depending on where the focus of analysis is placed. If the focus is on the population affected, the uncertainty is revealed as a "condition of waiting". When those who are in such situation are qualified professionals, what prevails may include official arguments questioned, and political debates which begin, and in some cases culminate, with appeals to judicial principles or challenging pollution parameters, such as the precautionary principle in the case of Ezeiza, or the acceptable thresholds for environmental pollution in the example of Colonia Hinojo. Contrary to the responses by professional groups, when working with more vulnerable communities, it can be observed that this condition of waiting becomes a "space of sacrifice". This can be reflected among different groups such as urban poor and indigenous or peasants who are exposed to recurrent dangers, and their situation worsens their difficulties to exercise rights in an effective way due to the lack of legal access to land tenure, among other causes, which limits them when it comes to prioritize their claims.

On the other hand, when one focuses on social actors, who implement technological innovations or public works, uncertainty may arise due to lack of information or contradictions among partial information supported by traditional beliefs and linked to the idealization of progress. This situation can generate false perceptions of safety as in the case of the constructions in La Boca neighborhood, or technological innovation in agriculture. Furthermore, when focus is centered on the institutions that manage risk for prevention or response, cultural devices are instruments of societies and politics to select information in the administration or management of catastrophes. We can see this in the case of the traditional institutionalization of memory and forgetting processes in a context of politicization of collective death and suffering, as the case studies on flooding in Santa Fe, and the "Cromañón" tragedy presented before.

Our work arises from a group of research articulated by the initiative of García-Acosta in the task of systematizing anthropological research of risk in Latin America. The challenges of this group consist of exploring different methodological approaches applied by anthropology to adjust a more appropriate method for risk analysis. This implies also that finding points of dialog is necessary with other disciplines and management approaches, which can reaffirm the contributions from anthropological analysis and the use of their own tools. The problems of risk, disasters, and crises require that social sciences not only enquire in a scientific academic terms, but also that they strike tensions from different forms to construct knowledge in order to identify uncertainty and therefore reduce it, which can be reflected in applied research.

Thus, from a complexity approach, we are developing new scientific academic research projects, as well as applied projects, focused on two lines: first, on social risk related to urban solid waste and, second, Mapuche and *criollo* communities living in oil industry areas, where unconventional gas is produced, at the Vaca Muerta deposit in Neuquén Province. This project focuses on understanding and strengthening local and disciplinary knowledge regarding vulnerability degrees in order to diagnose them in a collective way, and looks for conflict resolution mechanisms and risk reduction, involving affected communities, and public and private decision-makers from different administrative and jurisdictional levels.

Finally, we can assert that future investigations in Argentina could provide new contributions to the knowledge of risk studies from a social anthropological point of view. A focus on risk studies enables us to investigate topics which are of crucial importance in our discipline: the relationship between culture, risk, and social organization and the relationship between planning, meanings, and politics and its effects on affected populations with their responses. Nevertheless, we ought to search for new tools for generating new understandings in a fast changing society.

References

Abraham, E. (2009) Overview of the geography of the Monte Desert biome (Argentina). *Journal of Arid Environments*. 73 (2), pp. 144–153.

Abrutzky, R., Dawidowski, L., Murgida, A. & Natenzon, C. (2014) Contaminación del aire en la Ciudad Autónoma de Buenos Aires: El riesgo de hoy o el cambio climático futuro, una falsa opción. *Ciencia & Saúde Coletiva*. 19 (9), pp. 3763–3773.

Auyero, J. & Swistun, D. (2007) Expuestos y confundidos Un relato etnográfico sobre sufrimiento ambiental. *Iconos. Revista de Ciencias Sociales*. 28, pp. 137–152.

Baez Ullberg, S. (2013) *Watermarks: Urban flooding and memoryscape in Argentina*. Stockholm Studies in Social Anthropology N.S. 8. Stockholm, Acta Universitatis Stockholmiensis.

Baez Ullberg, S. (2015) La gestión de las inundaciones y la lógica de la omisión en la ciudad de Santa Fe. In: Viand, J. & Briones, F. (comps.). *Riesgos al Sur. Diversidad de riesgos de desastres en Argentina*. Buenos Aires, Imago Mundi Editores, LA RED, pp. 49–59.

Baez Ullberg, S. (2017a) La Contribución de la Antropología al Estudio de Crisis y Desastres en América Latina. *Iberoamericana-Nordic Journal of Latin American and Caribbean Studies*. 46 (1), pp. 1–5.

Baez Ullberg, S. (2017b) Desastre y Memoria Material: La Inundación de 2003 en Santa Fe, Argentina. *Iberoamericana-Nordic Journal of Latin American and Caribbean Studies*. 46 (1), pp. 42–53.

Balazote, A. (2001) Desinversión y riesgo en las lagunas Encadenadas. *Cuadernos de Trabajo*. 19, pp. 25–39.

Balazote, A. (1997) *Agua que no has de beber . . .* V Congreso Argentino de Antropología Social. La Plata, Argentina, julio–agosto.

Balbi, F. (2012) La integración dinámica de las perspectivas nativas en la investigación etnográfica. *Intersecciones en Antropología*, 13, pp. 485–499.

Bartolomé, L. (1985) Estrategias adaptativas de los pobres urbanos: el efecto 'entrópico' de la relocalización compulsiva. In: Bartolomé, L. J. (comp.). *Relocalizados: antropología social de las poblaciones desplazadas*. Buenos Aires, Ediciones del IDES, vol. 3, pp. 25–48.

Boivin, M., Rosato, A. & Balbi, F. (2000) Incidencia del evento de inundación de 1982–83 sobre el asentamiento humano en el área de islas del departamento de Victoria, Entre Ríos. *Relaciones de la Sociedad Argentina de Antropología*. Vol. XXV, pp. 27–40.

Bórmida, M. (1970) Mito y cultura. Bases para una ciencia de la conciencia mítica y una etnología tautegórica. *Runa. Archivos para las Ciencias del Hombre*. 12 (1–2), pp. 15–44.

Borrero, L. (2001) *El poblamiento de la Patagonia*. Buenos Aires, Emecé Ed.

Briones de Lanata, C. & Olivera, M. (1989) Luces y penumbras: Impacto de la construcción de la represa hidroeléctrica de Piedra del Águila en la agrupación Mapuche Ancatruz. *Cuadernos de Antropología*. Universidad Nacional de Luján. 2 (3), pp. 25–42.

Caputo, M. G., Hardoy, J. &Herzer, H. (1985) In: Caputo, M. G., Herzer, H. & Morello, J. (coords.). *Desastres naturales y Sociedad en América Latina*. Buenos Aires, Grupo Editor Latinoamericano, CLACSO. Available from http://65.182.2.242/docum/crid/Nov-Dic2003/pdf/spa/doc1012/doc1012-a.pdf. Accessed June 14th 2019.

Catullo, M. (1986) Relocalizaciones compulsivas de población. Estudio de un caso: ciudad Nueva Federación (Entre Ríos). *Runa*. Nueva Serie. XVI, pp. 137–156.

Catullo, M. & Brites, W. F. (2014) Procesos de Relocalizaciones. Las especificidades de los reasentamientos urbanos y su incidencia en las estrategias adaptativas. *Avá. Revista de Antropología social*. 25. Available from www.ava.unam.edu.ar/index.php/ava-25. Accessed June 14th 2019.

Díaz-Crovetto, G. (2015) Antropología y Catástrofes: intersecciones posibles a partir del caso Chaitén. *Revista Justiça do Direito*. 29 (1), pp. 131–144.

Etchichury, L., Gatti, I., D'Fabio, L., Murgida, A., Correa, M., Fontenla, L. & Membribe, A. (2016) Eventos extremos y riesgos. Diferencias y similitudes en políticas de gestión local del riesgo. Casos de Ingeniero Jacobacci, Neuquén, La Plata y Quilmes. In: Silva, M., Pérez, G. & Higuera, L. (ed.). *Geografías por venir*. Neuquén, Argentina, EDUCO, Universidad Nacional del Comahue, pp. 1309–1327. Available from www.researchgate. net/publication/316629000_eventos_extremos_y_riesgos_diferencias_y_similitudes_ en_politicas_de_gestion_local_del_riesgo_casos_de_ingeniero_jacobacci_neuquen_ la_plata_y_quilmes. Accessed June 14th 2019.

Funtowicz, S. & Ravetz, J. (1993) Science for the Post-Normal Age. *Futures*. 25 (7), pp. 739–755.

García-Acosta, V. (2018b) Vulnerabilidad y desastres. Génesis y alcances de una visión alternativa. In: González de la Rocha, M. & Saraví, G. A. (coord.). *Pobreza y Vulnerabilidad: debates y estudios contemporáneos en México*. Mexico, CIESAS, pp. 212–239.

García-Acosta, V. (2006) Estrategias adaptativas y amenazas climáticas. In: Urbina, J. & Martínez, J. (comp.). *Más allá del Cambio Climático: las dimensiones psicosociales del cambio ambiental global*. Mexico, Instituto Nacional de Ecología, Universidad Nacional Autónoma de México, pp. 29–46.

García-Acosta, V. (2005) El riesgo como construcción social y la construcción social de riesgos. *Desacatos. Revista de Antropología Social*. 19, pp. 11–24.

García-Acosta, V. (2004) La perspectiva histórica en la Antropología del riesgo y del desastre. Acercamientos metodológicos. *Relaciones. Estudios de historia y sociedad*. XXV (97), pp. 123–142.

García-Acosta, V. & Suárez, G. (1996) *Los sismos en la historia de México*, vol. I. Mexico, Fondo de Cultura Económica, CIESAS, Universidad Nacional Autónoma de México.

Gentile, E. E. (1993) *Estudio de la posible vinculación entre el fenómeno de El Niño-Oscilación del Sur y las crecidas extraordinarias del Paraná en su curso medio*. BA Thesis in Geography. Buenos Aires, Departamento de Geografía, Facultad de Filosofía y Letras, Universidad de Buenos Aires.

Girado, A. & Iturralde, R. S. (2015) Un abordaje socioantropológico del desastre ambiental y la percepción del riesgo en la Pampa Húmeda. In: Viand, J. & Briones, F. (comp.). *Riesgos al Sur. Diversidad de riesgos de desastres en Argentina*. Buenos Aires, Imago Mundi Editores, LA RED, pp. 125–141.

González, S. (1997) *Gestión urbana pública y desastres. Inundaciones en la baja cuenca del arroyo Maldonado(Capital Federal, 1880–1945)*. BA Thesis in Geography. Buenos Aires, Departamento de Geografía, Facultad de Filosofía y Letras, Universidad de Buenos Aires.

Guber, R. (2018) Anthropology in Argentina. In: Callan, H. (ed.). *The International Encyclopedia of Anthropology*. New York, John Wiley & Sons, Ltd., pp. 130–138. Available from https://doi.org/10.1002/9781118924396.wbiea1827. Accessed October 31st 2019.

Hansen, A. & Oliver-Smith, A. (1982) *Involuntary Migration and Resettlement: The Problems and Responses of Dislocated People*. Boulder, Westview Press.

Hermitte, E. & Bartolomé, L. (comp.). (1977) *Procesos de articulación social*. Buenos Aires, CLACSO, Amorrortu.

Herzer, H. (1993) *Catástrofes. Documentos de Base*. Seminario-taller La Universidad de Buenos Aires y el Medio Ambiente, Elementos para la Formulación de Políticas. Buenos Aires, Facultad de Filosofía y Letras, Universidad de Buenos Aires, May 26–28.

Herzer, H. (1990) Los desastres no son tan naturales como parecen. *Medio Ambiente y Urbanización*. 30, pp. 3–10.

Herzer, H. (1985) La inundación en el Gran Resistencia (provincia del Chaco, Argentina) 1982–1983. In: Caputo, M. G., Herzer, H. & Morello, J. (coords.). *Desastres naturales y Sociedad en América Latina*. Buenos Aires, Grupo Editor Latinoamericano, CLACSO.

Herzer, H. & Federovisky, S. (1989) Algunas conclusiones a partir de tres casos de inundaciones. *Medio Ambiente y Urbanización*. 26, pp. 18–24.

Hoffman, S. M. & Oliver-Smith, A. (ed.). (2002) *Catastrophe & Culture: The Anthropology of Disaster*. Santa Fe, Nuevo México, School of American Research.

Iturralde, R. (2015) Sufrimiento y riesgo ambiental. Un estudio de caso sobre las percepciones sociales de los vecinos de 30 de Agosto en el contexto de un conflicto socioambiental.

Cuadernos de Antropología Social. 41, pp. 79–92. Available from http://revistascientifi cas.filo.uba.ar/index.php/CAS/article/view/1597. Accessed June 14th 2019.

Lavell, A. (2005) Los conceptos, estudios y la práctica en el tema de los riesgos y desastres en América latina: evolución y cambio, 1980–2004: El rol de LA RED, sus miembros e instituciones de apoyo. In: *La gobernabilidad en América Latina*. San Jose de Costa Rica, Secretaría General FLACSO, pp. 2–66. Available from http://biblioteca virtual. clacso.org.ar/ar/libros/flacso/secgen/Lavell.pdf. Accessed June 14th 2019.

Lavell, A. (1996) Degradación ambiental, riesgo y desastre urbano. problemas y conceptos: hacia la definición de una agenda de investigación. In: Fernández, M. A. (comp.). *Ciudades en riesgo. Degradación ambiental, riesgos urbanos y desastres en América Latina*. Lima, LA RED, USAID. Available from http://www.desenredando.org/public/ libros/1996/cer/CER_cap02-DARDU_ene-7-2003.pdf. Accessed June 14th 2019

Maskrey, A. (comp.). (1993) *Los desastres no son naturales*. Bogota, LA RED, Tercer Mundo Editores.

Mastrángelo, A. (2000) Londres y Catamarca. La articulación rural/urbano en una localidad del N.O. argentino a finales del siglo XX. *Horizontes Antropológicos*. 6 (13), pp. 89–112.

Merlinsky, G. (2017) Cartografías del conflicto ambiental en Argentina. Notas teóricometodológicas. *Acta sociológica*. 73, pp. 221–246.

Merlinsky, G. (2016) Mists of the Riachuelo: River Basins and Climate Change in Buenos Aires. *Latin American Perspectives*. 43, pp. 43–55.

Merlinsky, G. (2013) *Política, derechos y justicia ambiental. El conflicto del Riachuelo*. Buenos Aires, Fondo de Cultura Económica.

Moreno, L. (2015) Riesgo y políticas públicas: disputas en el proceso de urbanización de una "villa de emergencia" en la cuenca del río Reconquista. In: Viand, J. & Briones, F. (comp.). *Riesgos al Sur. Diversidad de riesgos de desastres en Argentina*. Buenos Aires, Imago Mundi Editores, LA RED, pp. 21–34.

Murgida, A. (2013) Cambios socio-ambientales: desplazamientos de las poblaciones históricamente postergadas en el chaco-salteño. *Cuadernos de Antropología*. 9, pp. 35–63.

Murgida, A. (2012) *Dinámica Climática, Vulnerabilidad y Riesgo. Valoraciones y procesos adaptativos en un estudio de caso del Chaco-salteño*. PhD Thesis in Anthropological Sciences. Buenos Aires, Facultad de Filosofía y Letras, Universidad de Buenos Aires.

Murgida, A. (2000) *Riesgo e identidad en la protesta social local: el caso de J. Ma. Ezeiza*. BA Thesis in Anthropological Sciences. Buenos Aires, Facultad de Filosofía y Letras, Universidad de Buenos Aires.

Murgida, A. & Gasparotto, M. (2015) Percepción del riesgo y sistemas participativos de alerta temprano en Iruya, Provincia de Salta. In: Natenzon, C. & Ríos, D. (comp.). *Riesgo, catástrofe y vulnerabilidad. Aportes desde la Geografía y las ciencias sociales para casos argentinos*. Buenos Aires, Editorial Imago Mundi, pp. 74–95.

Murgida, A. & Gentile, E. E. (2015) Aceptabilidad y amplificación del riesgo en la estepa nor-patagónica. In: Viand, J. & Briones, F. (comps.). *Riesgos al Sur. Diversidad de riesgos de desastres en Argentina*. Buenos Aires, Imago Mundi Editores, LA RED, pp. 195–213.

Murgida, A. & González, S. (2004) *Social Risk, Climate Change and Human Security: An Introductory Study Case in Metropolitan Area of Buenos Aires. Argentina*. Proceedings of International Workshop: Human Security and Climate Change, Oslo, June 21.

Murgida, A., González, M. & Tiessen, H. (2014) Rainfall Trends, Land Use and Adaptation in the Chaco Salteño Region of Argentina. *Journal of Climatic Change*. 14 (4), pp. 1387–1394. Available from http://dx.doi.org/10.1007/s10113-013-0581-9. Accessed June 14th 2019.

Murgida, A., Guebel, C., Natenzon, C. & Frasco, L. (2013) El aire en la agenda pública de la Ciudad Autónoma de Buenos Aires. In: Sánchez Rodríguez, R. (ed.). *Respuestas Urbanas para el Cambio Climático*. Santiago, IAI-CEPAL-UN, pp. 137–157.

Murgida, A. & Kazimierski, M. (2017) Proyectos de interfaz ciencia – política y la reducción de incertidumbre en el desarrollo productivo en el Comahue. In: Pérez Carrera, A. & Volpedo, A. (comp.). *El desarrollo agropecuario argentino en el contexto del cambio climático: una mirada desde el PIUBACC*. Buenos Aires, EUDEBA, pp. 75–87.

Murgida, A., Laham, M., Chiappe, C. & Kazimierski, M. (2017) Modos adaptativos bajo condiciones hidroclimáticas extremas. In: Pérez Carrera, A. & Volpedo, A. (comp.). *El desarrollo agropecuario argentino en el contexto del cambio climático: una mirada desde el PIUBACC*. Buenos Aires, EUDEBA, pp. 65–74.

Murgida, A., Laham, M., Chiappe, C. & Kazimierski, M. (2016) Desarrollo social bajo sequía y cenizas. *Iluminuras*. 17 (41), pp. 11–29.

Murgida, A. & Natenzon, C. (2009) Social Downscaling: A Few Reflections on Adaptation in Urban Environments. In: Leite da Silva Dias, P. et al. (ed.). *Public Policy, Mitigation and Climate Change in South America*. São Paulo, Instituto de Estudos Avançados da Universidade de São Paulo, pp. 139–156. Available from www.iea.usp.br/publi cacoes/textos/copy_of_climatechangeandsouthamerica.pdf/at_download/file. Accessed June 14th 2019.

Natenzon, C. (2015) Presentación. In: Natenzon, C. & Ríos, D. (comp.). *Riesgo, catástrofe y vulnerabilidad. Aportes desde la Geografía y las ciencias sociales para casos argentinos*. Buenos Aires, Editorial Imago Mundi, pp. X–XXV.

Natenzon, C. (2003) *Inundaciones catastróficas, vulnerabilidad social y adaptaciones en un caso argentino actual. Cambio climático, elevación del nivel medio del mar y sus implicancias*. Forum Workshop IX Climate Change Impacts and Integrated Assessment Energy Modeling, Stanford University, July 28–August 7.

Natenzon, C. (1995) *Catástrofes naturales, riesgo e incertidumbre*. Argentina, FLACSO, Serie Documentos e Informes de Investigación Num. 197.

Oliver-Smith, A (2002) Theorizing Disasters: Nature, Power, and Culture. In: Hoffman, S. M. & Oliver-Smith, A. (ed.). *Catastrophe & Culture: The Anthropology of Disaster*. Santa Fe, Nuevo México, School of American Research, pp. 23–47.

Poggiese, H. & Francioni, M. (1993) *Escenarios de gestión asociada y nuevas fronteras entre el Estado y la sociedad*. Second International Conference, International Institute of Administrative Sciences, Toluca.

Radovich, J. (2013) Las Ciencias Sociales y los procesos catastróficos. Aspectos teórico/metodológicos y estudios de caso: las erupciones volcánicas en Patagonia en años recientes. In: Balazote, A. & Radovich, J. (comp.). *Estudios de Antropología Rural*. Buenos Aires, Editorial de la Facultad de Filosofía y Letras, Universidad de Buenos Aires, pp. 21–50.

Radovich, J. & Balazote, A. (1998) *Impacto social de grandes obras de infraestructura en la agrupación mapuche Painemil, Provincia de Neuquén*. Informe final, programación UBACyT (FI 158) 1995–97. Sección Antropología social. Buenos Aires, Instituto de Ciencias Antropológicas, Facultad de Filosofía y Letras, Universidad de Buenos Aires.

Radovich, J. & Balazote, A. (1996) Inversión y desinversión de capital en megaproyectos hidroenergéticos. Efectos sociales en poblaciones mapuche asentadas sobre los ríos Limay y Neuquen. *Papeles de Trabajo*. Rosario, Centro Interdisciplinario de

Ciencias Etnolingüísticas y Antropológico Sociales, Universidad Nacional de Rosario. 6, pp. 12–27.

Ratier, H. (1971) *El cabecita Negra.* Buenos Aires, Centro Editor de América Latina. Colección "La historia popular" Num. 72.

Ribeiro, G. L. (1985) Proyectos de Gran Escala: Hacia un marco conceptual para el análisis de una forma de producción temporaria. In: Bartolomé, L. (comp.). *Relocalizados: Antropología Social de las poblaciones desplazadas.* Buenos Aires, Ed. IDES, Vol. 3, pp. 23–47.

Riera, C. (2017) La tecnología de riego y la disputa por el agua subterránea en Córdoba, Argentina. *Caderno de Geografia.* 27 (48), pp. 27–43.

Riera, C. & Pereira, S. (2015) Vulnerabilidades e incertidumbres de la innovación tecnológica en la agricultura bajo riego en la provincia de Córdoba. In: Natenzon, C. & Ríos, D. (comp.). *Riesgo, catástrofe y vulnerabilidad. Aportes desde la Geografía y las ciencias sociales para casos argentinos.* Buenos Aires, Editorial Imago Mundi, pp. 53–74.

Ríos, D. & Murgida, A. (2004) Vulnerabilidad Cultural y Escenarios de Riesgo por Inundaciones. *GEOUSP Espaço e Tempo.* 16, pp. 181–193.

Roze, J. (1997) *Más que pobres: pobreza y estigmatización: los inundados de Resistencia.* I Congreso Internacional Pobres y Pobreza en la Sociedad Argentina. Universidad Nacional de Quilmes. Available from www.equiponaya.com.ar/congresos/contenido/quilmes/P3/28.htm. Accessed June 5th 2019.

Sarlingo, M. (1995) *La ciudad fragmentada.* Olavarría, Instituto de Investigaciones Antropológicas.

Suárez, F. (1994) Con el corazón en la boca: las metáforas de una inundación. *Desastres & Sociedad.* 3 (2), pp. 39–45.

Swistun, D. (2015) Desastres en cámara lenta: incubación de confusión tóxica y emergencia de justicia ambiental y ciudadanía biológica. *O Social em Questão.* XVIII (33), pp. 193–214.

Valiente, S. & Radovich, J. (2016) Disputas en el territorio por actividades tipo enclave en Norpatagonia y Patagonia Austral, Argentina. *Cardinalis.* 4, pp. 35–67.

Visacovsky, S. (2017) When Time Freezes: Socio-Anthropological Research on Social Crises. *Iberoamericana-Nordic Journal of Latin American and Caribbean Studies.* 46 (1), pp. 6–16.

Warner, K., Hamza, M., Oliver-Smith, A., Renaud, F. & Julca, A. (2010) Climate Change, Environmental Degradation and Migration. *Natural Hazards.* 55, pp. 689–715.

Zenobi, D. (2017) Políticas Para La Tragedia: Estado y Expertos en Situaciones de crisis. *Iberoamericana-Nordic Journal of Latin American and Caribbean Studies,* 46 (1), pp. 30–41.

Zenobi, D. (2015) Del incendio al santuario de Cromañón. Notas sobre un proceso social. In: Viand, J. & Briones, F. (comp.). *Riesgos al Sur. Diversidad de riesgos de desastres en Argentina.* Buenos Aires, Imago Mundi Editores, LA RED, pp. 35–46.

2 The field of Anthropology of Disasters in Brazil

Challenges and perspectives

Renzo Taddei

Introduction[1]

Brazil has a well-developed and vibrant anthropological community. In 2016, the country had 49 graduate programs in the discipline, distributed among 29 university departments of anthropology or social sciences. According to official records of the Coordination for the Improvement of Higher Education Personnel (CAPES), the agency of the national Ministry of Education dedicated to postgraduate programs, a growth of 300% took place in 16 years (2000–2016) in a number of graduate courses.

It also has continental territorial dimensions, with a great diversity of ecosystems and biomes. It adopted a politico-economic model that combines capitalistic extractivism and a 19th-century ideology of modernization that prescribes heavy-handed human domination over nature. Moreover, it is one of the most socioeconomically unequal countries on the planet. The conjunction of these three factors generates conditions of vulnerability of a different sort and produce disasters of all types.

In spite of the vitality of Brazilian anthropology and of the frequency with which disasters hit Brazilian populations, the field of Anthropology of Disasters in the country is not formalized. This chapter intends to explore the reasons for such a state of affairs, and analyzes recent transformations in the recent Brazilian anthropological panorama in an attempt to forecast the future of this field of research in the country.

The main argument of this chapter is organized around two facts: the first is that, despite the systematic historical occurrence of events that typically could be considered as being "disasters", subsisted in the collective imagination of mainstream society, throughout the 20th century and in the early years of the 21st, the idea that "there are no disasters" in the country. The second is that this fact seems to be undergoing transformation, due to "natural" and "technological" disasters that hit the political and economic centers of the country in the last two decades. Anthropological agendas follow this same path. The underlying questions that deserve exploration, and that will tangentially be addressed in this text, refers to what the conditions and processes are that turn something in the world into an object of anthropological treatment, and what relation this has with

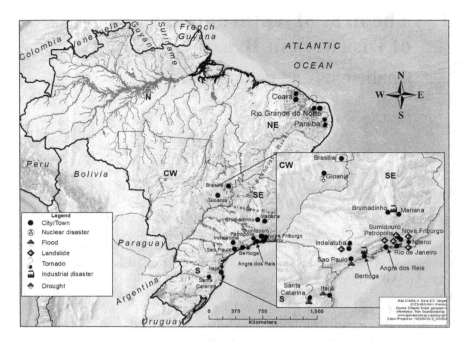

Map 2.1 Map Brazil. Case studies and main areas mentioned

transformations in patterns of mainstream collective imagination. In a word, disasters are imagined, and are also de-imagined, and both alternatives have cultural and political consequences.

The text was deliberately written in the form of (speculative) essay; it is not a chronology of works and authors, although reference to main bibliographic contributions are provided. The categories used for making reference to the geographic divisions of Brazil are the ones adopted by the Brazilian Institute for Geography and Statistics (IBGE), which are widely adopted in the country, including by scholars in their attempt to address historical political and economic inequalities in relation to disasters (see, for instance, Albuquerque Jr., 2014).

Brazil's invisible disasters

Let me state this fact right from the outset: there is (almost) no disaster in Brazilian anthropology. A brief evaluation of the seven journals edited in Brazil or in the Portuguese language,[2] with highest impact factor,[3] carried out in November 2013, demonstrates that out of 187 editions and over 1,300 articles, all open access at the Scientific Electronic Library Online (Scielo), a search using the keywords *disaster, tragedy, risk, vulnerability, resilience*, and *climate* resulted in only 14 articles, or 1% of the available texts. The keywords *disaster, tragedy* (in a nonliterary sense), *vulnerability*, and *resilience* did not produce one single result. Of the 14 found articles, 13 deal with the concept of risk and one with climate change. The search was repeated in 2017, with no qualitative change in its results. Other methods for data generation would perhaps present different results; still, even if the method implied here is not appropriate to characterize *the whole field* of anthropology in Brazil, it has the virtue of focusing on the journals and texts that have the capacity for setting agendas and tendencies inside Brazil's anthropological community.

There are three important hypotheses that may help in the effort of making sense of such a state of affairs. The first one refers to a mismatch of categories: an objection could be raised that the absence of keywords associated to the concept of disaster in the bibliography is due the fact that it is a Western category. The attention put on *emic* categories in ethnographic work may have fooled search engines such as the one used by the Scielo platform. It does not seem to me that this could be the case, though. Throughout the 20th century, Brazilian anthropology was not less colonial and patronizing to indigenous ideas than its American and European counterparts, as indigenous intellectuals from Brazil and abroad are quick to point out (Deloria Jr., 1969; Baniwa, 2016); the general tendency was that emic categories were quickly dissolved into Western universalisms.[4] Additionally, urban anthropology is as robust in Brazil as is indigenous ethnology, and the topic of disasters has been absent there too. We can, therefore, discard the option of categorial confusion.

A second possibility refers to specificities of the intestine topographies of Brazilian anthropology; or, to put it more directly, the possibility that what is framed as "disaster" in certain approaches and traditions may be framed otherwise in Brazilian academia. There is a great amount of evidence that this indeed explains part of the paradox. A number of important anthropologists in Brazil refuses theoretical trends that address disasters as "excess", that is, as phenomena that systematically expose the limits of our conceptual schemes. The work of some of the anthropologists who have published on mining-related catastrophes call attention to the fact that the very nomenclature, "technological disaster", is an expression that carries in itself a political risk. The idea of "normal accidents", for instance, proposed by Charles Perrow in 1984, while conceptually important, is at the same time politically dangerous in how it depoliticizes the phenomena it describes. In the case of what is perhaps the two worst technological disasters in the history of the country, the rupture of mining ore reject dams in two localities in the state of Minas Gerais, with a little more than three years between the two events – the "disasters of Mariana and Brumadinho" (to be presented in more detail later) – the existence of hard evidence that the management of the mining companies (Samarco and Vale) knew about the bad conditions of the dams led some authors to denounce the term "disaster" and to advocate for the use of the term "crime" instead (Reis & Santos, 2017; Zhouri et al., 2016a, 2016b, 2017).

By extension, something similar could be argued in relation to the effort of institutionalizing the Anthropology of Disasters as a subfield of Brazilian anthropology: what is gained and what is lost, one could ask, in reframing a research that has been historically understood as Environmental Anthropology in a context of ecological degradation by the mining industry, or research on/ in the situations of political conflicts in which traditional populations (peasants, fisherpersons, indigenous communities) have their rights expropriated by extractivist corporations, when and if these research efforts become "anthropologies of disasters"? Part of the concern is associated to the construction of equivalences between the academic research and the world of public policy/politics, so as to make research more politically effective. While works framed as Environmental Anthropology may find interlocution with existing environmental agencies in the body of the state, at municipal and state levels, the issue of "disaster" is largely absent from the state apparatus, with the exception of very centralized and highly inaccessible agencies at the federal government, or the highly militarized civil defense apparatus (Valencio, 2009, 2010). It is very clear that, from the perspective of the work with impacted populations, the political economy of categories is dramatically important, and it may enhance, or dilute, the political efficacy of the work with the communities. As all disciplinary fields are subject to similar performative effects of labels and divisions (Bourdieu, 1979), this reflection itself is no stranger to the discipline.

All that said, it is important to notice that the category of "disaster" is extremely fluid, and extends to diverse territories, populations, and circumstances, many of which are disconnected to contexts in which political struggle and advocacy opportunities are clearly identifiable (for complex ecopolitical reasons), even if they exist. Tornados in southern Brazil and droughts in the Amazon region are cases in point. So while all elements of the argument presented here are still in place, and phenomena that in other contexts would be framed as disasters exist in Brazil with different conceptual make ups, there is a great amount of "disastrous" phenomena that are simply never anthropologized, no matter what.

That brings us to the third hypothesis to be considered in the effort to make sense of the absence of disasters in anthropological production: this fact reflects another absence, of larger demographic amplitude – the widely disseminated idea, in effect throughout the 20th century in Brazil, that the incidence of disasters in the country was minimal, if not completely absent. The perception that the country was awarded with a "benign nature", from which the "no-disaster" idea seems to spring, is deeply rooted in the collective imagination of the country; it probably reflects some elements of the European imagination about the America's during early colonial times (Wasserman, 1994: p. 30). One example from popular culture is an old joke, still in circulation, in which Gabriel, the angel, asks God why he spared Brazil from natural disasters when creating the country, and God answers that the people he would put there would be a disaster in itself (Strasdas, 2011).[5] Machado, in an article about one the many 19th-century European scientific expeditions in Brazil, writes:

> To a Brazilian, nothing sounds more familiar than statements that Brazil has been blessed in terms of nature. Indeed, from our earliest childhood, we learn

to identify our country through enthusiastic manifestations about the wonders of our geography, not to mention the flora and fauna whose extraordinary diversity and wealth comprise the treasures that God generously bestowed upon us. . . . Pero Vaz de Caminha's inspired letter reporting the discovery of new lands to the King of Portugal in 1500, . . . described these lands in term of Eden.

(Machado, 2004: p. 13)

Against all that, a consultation to news archives promptly provides evidence of the cyclical occurrence of epidemies of suffering caused by extreme events, of environmental and/or technological nature, in the country: droughts (with the frequency of one in every five years in the Northeast region; less frequent in other parts of the country, but still prevalent in the Amazon and the southern regions), earthquakes in the Northeast region, destructive floods in the Amazon, and tornados and floods accompanied by landslides in the southeastern and southern regions – just to mention a few. It is remarkable that the benign nature narrative, perhaps the most prevalent myth of origin of the Brazilian nation, had the power to obfuscate the cyclicity and systematicity of the epidemies of suffering caused by disasters, leaving no strong mark in the collective imagination; or at least not in the collective imaginary of the social groups with power and capacity to generate and disseminate narratives about Brazil, such as the cultural industries of the southeastern region (at cities such as São Paulo and Rio de Janeiro), and the public policies created at the national capital, Brasilia.[6] It is precisely the recent occurrence of large-scale disasters in the southeastern region – a contingency of the destiny, or the effects of climate change, or both – what seems to be changing the picture.

Brazilian social science's invisible disasters

It is necessary to note that disasters are always thought through politics (Oliver-Smith, 2010), and the political history of Brazil cannot be thought as detached from ideologies and projects of modernization (Taddei & Gamboggi, 2010, 2011). Disasters, in such context, play the double role of being agents of marginalization, while being marginalized, at the same time. In the case of droughts, for instance, they were, and still are, often understood as an impediment for progress; that is made evident in the very name of the oldest federal agency created to handle natural disasters in Brazil, the National Department for Works Against the Drought (DNOCS) – note the *against*. Droughts are also accused of being responsible for the "backwardness" of the semiarid Northeast region. The disasters that took place in the richest and most powerful region of the country, the Southeast, were treated as transient, episodic phenomena, which is what generated the perception that the economic centers of the country have been mildly affected by droughts and floods historically, if compared to marginal areas such as the Northeast and the Amazon. In a typical chicken and egg situation, in this teleological perceptive scheme, power and progress seem to minimize disasters, and these in turn are thought to have spared power and progress, reinforcing the dominance of the richest region of Brazil, the Southeast, as if the sociopolitical inequalities were a "natural" (when not theological) phenomena. One way of understanding such a complex web of relations is that disasters were historically constructed as natural-political devices for "naturalizing" political and economic inequality (Oliver-Smith & Hoffman, 1999; Hoffman & Oliver-Smith, 2002).

A somewhat parallel strand of events took place in the history of the social sciences in Brazil. Despite of the importance of Northeast-based authors like Gilberto Freyre and Câmara Cascudo,[7] the institutionalization of the social sciences in Brazil occurs in the Southeast. Important historical moments were the creation of the Faculty of Philosophy, Letters, and Human Sciences (FFLCH) at the University of São Paulo (USP) – with the participation of Claude Levi-Strauss – from which emerged the Paulista Sociological School; the foundation of the Brazilian Anthropological Association (ABA), in 1955; and the creation of the first graduate program in Anthropology at the National Museum, in 1968, in Rio de Janeiro.[8]

Even if social scientists in São Paulo and Rio de Janeiro were interested in what happened in the rest of Brazil – the centrality of indigenous ethnology in the academic production of the era demonstrates that – the research agenda had as a wider ideological context the question of modernism, modernization, and the construction of the Brazilian civilization. As it is widely known, in a typically modern understanding of reality, nature is to be dominated and exploited. It was left to anthropologists to document and make intellectual sense of the reality of the victims of modernization: the context in which disasters gained the form of genocide of indigenous and traditional populations. These genocides, nevertheless, took part in the political order in place, a political order that, despite of being extremely perverse, had its intrinsic logic. In the context of origin and initial developments of social sciences in Brazil, the country was thought from, and through, the modernizing Southeast. "Total" disasters remained beyond cognition, and therefore remained invisible.[9]

One theoretical aspect of the problem being addressed here lies exactly in the aforementioned question of the reliance on a metaphysical assumption of existence of order. Brazilian intellectuals were, unsurprisingly, mimicking European colleagues here. Both Émile Durkheim's *social fact* (1982 [1895]) and Max Weber's *ideal type* (1949 [1904]) imply the effort in understating societies in their normal, typical, and ordinary conditions. These ideas induced researchers to discard the extraordinary as irrelevant – even if the kernel of the question lies in understanding what exactly counts as extraordinary, *and to whom*. Evans-Pritchard (1940), for instance, describes the typical organization of the Nuer and from it he erases the influence of English colonialism in Sudan; likewise, occurrences that generate the radical disorganization in studied social forms are abstracted or sent to the background of social action, so as to stress what is supposedly most relevant in the theoretical realm: it is not the drought (something that can cause anomie) that interests the researcher, but the processes of accusation of witchcraft (as a sociologically established form of reproducing social order) that are associated to it (Evans-Pritchard, 1976), and so on.

Brazilian disasters made visible

In the last two decades, a series of facts, in three distinct fronts, seem to have started a process of reversion of the scenario described above. The three fronts are these: a) the occurrence of disasters of great proportions in the Southeast region of Brazil, what generated a process of transformation in the configuration of state agencies dedicated to disaster prevention and relief in the country; b) the occurrence of international disasters that affected Brazil in inedited forms; and c) a series of new developments in social theory that place the issue of disasters in new analytical keys.

Disasters under the (national) spot

In the first of the aforementioned fronts, Brazil has some of the most vulnerable ecosystems to climate change (rain forest, semiarid Northeast, and the savannas of the center-west), which, when added to environmental devastation, population growth, and poorly managed urban expansion, creates the appropriate conditions for an increase in number of disasters. And indeed, the last 20 years witnessed a significant increase in the frequency of highly visible disasters. And yet, as expected, this process, still in course, does not take place without some amount of epistemological conflict. Two significant examples are related to hurricanes and tornados. In March of 2004, hurricane Catarina made landfall in the state of Santa Catarina (Lopes, 2015; Klanovicz, 2010). Some scientists argued that Catarina could not be a hurricane, due to the simple fact that "there are no hurricanes in Brazil".[10] The same happened with tornados: in the years of 2002 and 2003, while I conducted ethnographic fieldwork in the state of Ceará, I documented in interviews narratives on strong snorters, locally called "ventanias", in the metropolitan region of Fortaleza and in the Apodi hills. The described linear pattern of destruction immediately called my attention; when I inquired about whether those winds were not in fact tornados, I heard from farmers and from meteorologists that they believed so, but they could not openly talk about it, out of fear of being ridiculed. In a way, seeing a tornado in Brazil at that time was equivalent to seeing a ghost. It was with the tornado in the central area of the city of Indaiatuba, in the state of São Paulo, in 2005, that Brazil discovered that there are tornados in national territory; there is even a "tornado alley" in the country (Candido, 2012), and Brazil seems to have the second highest frequency of tornados on the planet (Catucci, 2012).

Something similar happens with earthquakes, which occur at an average rate of over a thousand per year in the states of Ceará, Rio Grande do Norte, and Paraíba, with intensity that tend not to surpass level 3 on the Richter scale (Moreira, 2013).[11]

Droughts occupy a special and peculiar place in the collective imagination of disasters in Brazil: of all phenomena that are typically classified as disasters, none is more documented and studied in the country than droughts (see, for instance, Gareis et al., 1997; Kenny, 2002, 2009; Nelson & Finan, 2009; Palacios, 1996; Pennesi, 2013; Taddei, 2012, 2013). And yet, the prevalence of droughts puts into evidence the bizarre contours of mental maps: if asked about disasters, most

Brazilians would say the country is blessed with their absence; if asked about *droughts*, the same people would say that they are "typically" northeasterner. And "typical" northeasterner droughts are accompanied with "typical" images of suffering and hunger – it is not the case of the (Brazilian imagination of) "Israeli" droughts, the ones that occur and no one notices their presence.[12] The drought is, then, the most common disaster in a country devoid of disasters.

And again, atypical events seem to have happened in the recent past: droughts crossed geographic barriers, and also political and imaginary barriers in result. In 2005, Brazil has witnessed droughts of large proportions in three of its main regions at the same time: the Amazon, the Northeast region, and the southern region (Taddei & Gamboggi, 2010). The concurrence of the three droughts contradicts the widespread idea that the El Niño phenomenon induces extreme events of alternated nature in the country: droughts in the Northeast occur in parallel with floods in the southern region. The drought at the Amazon basin was so intense, and its impacts so dramatic to the ecosystem and the communities, that the event quickly gained the attention of the international press (e.g., Rohter, 2005). Devastating droughts returned to the Northeast in subsequent years: in 2007, in 2010, and finally between 2011 and 2018, in what is considered the longest drought ever recorded for the region. In 2010 the Amazon basin was also hit by what was considered the worst drought in the last 100 years. Then, between 2012 and 2016, the largest metropolis of the country, the city of São Paulo, suffered the worst drought and water storage crisis in its history, with the depletion of one of its strategic reservoir complexes: the Cantareira system (Leite, 2018). "This is a situation we only saw in the Northeast, on TV", was a phrase frequently repeated in São Paulo, with a tone that suggested that the drought was being accused of geopolitical insolence.

Concomitantly, a number of disasters associated to excessive rains took place, generating floods and landslides, and mobilizing great media coverage and repercussions with public opinion, especially around those taking place in the Southeast region. In 2008, floods in the Itajaí valley, state of Santa Catarina, made over 130 fatal victims (Silva, 2013). The floods at the same valley happened again, on an equally destructive scale, in 2011 and 2013. In the first day of the year 2010, a large landslide in the city of Angra dos Reis, state of Rio de Janeiro, made 50 victims. In that same year another landslide in the state of Rio de Janeiro, this time in the city of Niterói, killed over 200 people. In the following year, a massive landslide in the Fluminense hills hit the municipalities of Nova Friburgo, Teresópolis, Petrópolis, Sumidouro, São José do Vale do Rio Preto, and Bom Jardim, killing 840 individuals and leaving 440 disappeared (Silva, 2015). It is relevant to the argument of this text that many of these cities provide mountain tourism attractions to wealthy inhabitants of Rio de Janeiro.

There were also technological disasters, especially oil spills, with its dramatic impacts to life and coastal ecosystems. In 2010, a spill occurred at the processing platform P-47, field of Marlim, in the Campos Basin (near the city of Macaé, Rio de Janeiro). In the following year, another spill occurred, this time at a Chevron platform, equally in the Campos Basin. In 2013, there was a spill in the city of Bertioga, in the state of São Paulo.

Added to all that was mentioned, the already-cited cases of the mining disaster-crimes of Mariana and Brumadinho infamously occupy a prominent role in the recent history of disasters (and of the anthropological analyses of them) in Brazil. An occurrence that took place initially in the community of Bento Rodrigues, in the municipality of Mariana, state of Minas Gerais, then possibly became the worst industrial-environmental tragedy in Brazilian history, worked as a catalyzer for the rapid growth of the group of anthropological professionals and students doing systematic and robust work on disasters,[13] – even if, again, the category of disaster is actively resisted. In the morning of 5 November 2015, a reservoir of rejects in an iron ore mine (owned by Samarco, which is controlled by Vale and the Anglo-Australian mining corporation BHP) broke, devastating a number of communities located downstream from it. The mud killed at least 19 people, directly displaced over 200 families; it then reached the Doce river, killing all forms of life in more than 663 kilometers of river, and then entering the sea, where it created a fan of devastation. All the riverine populations, including the Krenak Indians and a large number of traditional communities of fisherpersons, were impacted by the destruction of their main source of food and income. The city of Governador Valadares, with 280 thousand inhabitants, had the rio Doce as its main source of fresh water, and suffered a severe water shortage after the event. Thirty-nine cities are "officially" considered as affected by the disaster; the number of municipalities that rely on the Doce river for water is around 230.

Then, on 25 January 2019, another iron ore rejects reservoir (owned by Vale) broke, this time in the community of Córrego do Feijão, in the municipality of Brumadinho – less than 200 kilometers away from Mariana. Over 200 individuals were killed, and over 93 remain disappeared. Two months after the event, the wave of toxic mud, after having devastated over 200 kilometers of the ecosystem of the Paraopeba river and affecting hundreds of cities and tens of thousands of individuals, reached the São Francisco river, the fourth largest river in South America and the most important river outside of the Amazon basin.

A report by the National Water Agency in Brazil issued in November of 2018 announced that the country had 45 dams in critical condition; the one that broke in Brumadinho was not among them. The socioecological damage to the ecosystems of the areas of both occurrences (that together are almost the size of Portugal) does not fit in words and images; the population directly affected may well be in the level of millions of individuals.

This dismal accountancy doesn't have any other goal than to make evident that the convoluted existence of the elements is not aligned to the cultural representations, historically shaped, of nature that are placid and passive to the assaults of modernization. In any case, the occurrences here mentioned (and many others) in the last two decades have filled the national and local news, and seem to be slowly transforming the imaginary of the mainstream Brazilian population on what concerns "their" disasters.

The idea that a major change in how disasters are institutionally dealt with in Brazil would require the occurrence of disasters in the Southeast region seem to be confirmed by how the federal government responded to the massive landslide that happened in the Fluminense hills in 2011. In 2005, the National Center for

Management of Risks and Disasters (CENAD) was created, under the Ministry for National Integration, and installed in a small room with 30 square meters of area. In 2011, in the wake of the national commotion created by the mentioned landslide, it was restructured and transferred to a new space, 20 times bigger in area. CENAD works in partnership with the National Center for Disaster Monitoring and Alerts (CEMADEM), an agency which is linked to the Ministry of Science and Technology and also created in the same circumstance, in 2011.

Brazil in overseas disasters, and vice-versa

The participation of Brazilian citizens in international disaster events, a fact that has been widely exploited by the national media, is a second front in the transformation of perspectives on the national imagination of disasters. The economic stabilization and household income growth that characterized the 1994–2010 period in Brazil, added to developments in transportation and mobile communication technologies, placed Brazilians – tourists, professionals, and diplomats – in the frontlines of disasters as victims, but also as producers and disseminators of images and information. That happened in situations like the tsunami of the Indian Ocean, in 2004, when national media chains did not have difficulty in finding Brazilian citizens who had witnessed the disaster and produced photographic and video records of it. The media coverage focused, in large part, on the death of Brazilian diplomat Lys Amayo de Benedek D'Avola and her ten-year-old son. In the 2010 Haiti earthquake (Thomaz, 2010; Bersani, 2015), the strong presence of UN Brazilian troops in the country, and the death of Zilda Arns, head of the Catholic Pastoral Care for Poor Children, made the disaster one of the most intensely covered international events by the Brazilian media that year.

In what concerns technological disasters, also in 2010, the Deepwater Horizon/British Petroleum oil spill – probably the worst oil spill in history – took place, unequivocally alerting nations and companies about the perils of deep water oil prospection, in a historical moment in which Brazil was inebriated by the artificially inflated euphory of the discovery and beginning of operations of what became known as its "pre-salt" oil fields. In the following year, the Japanese tsunami and the ensuing nuclear disaster in Fukushima brought back old nightmares – not only those related to Three Mile Island and Chernobyl, but also of the case of Cesium 137 on Goiania (Silva, 2017; Fonseca & Klanovicz, 2014; Vieira, 2010, 2013; Queiroz, 2017), which occurred in 1987. Maps of nuclear accidents and disasters appeared once again on the Internet and on TV, and the Goiania accident reminded the Brazilian population that the country features in the list of nations where nuclear accidents occurred. New exercises of geographical imagination appeared after Fukushima: as American West Coast universities started detecting signs of radiation in salmon and tuna captured in the Pacific, a consequence of the uninterrupted flow of radioactive water from the damaged Fukushima Daiichi Nuclear Power Plant into the ocean, middle-class Brazilians became concerned with the source of the salmon they eat in their sushi and temaki.[14]

Social theory

The third front refers to transformations in the panorama of the contemporary social sciences. In the international English-speaking social sciences scene, sociology clearly moved ahead of anthropology; the same can be said to have happened in Brazil.[15] The theories of Ulrich Beck (1992) and Anthony Giddens (1990, 1991) on the *risk society* left an important mark in the sociological panorama: risk was elevated to a constitutive element of the very ontology of contemporaneity, while becoming, at the same time, a new metaphysics. Charles Perrow (1999), in his turn, proposed the idea of *normal accidents*, in which complex systems, technological or not, may assume internal configurations that, despite their catastrophic nature, are nothing more than one of its possible, and therefore "normal," states. A disaster, in result, is revealed as one of the possible states of reality.

In the border between sociology and anthropology, Bruno Latour (1991, 2013) and other colleagues working with the Actor-Network Theory (ANT) articulate a double accusation against the so-called hard sciences and also against the so-called social sciences. While the "natural" or "formalist" science is evoked to disarticulate politics, that is, everything else of which it is not composed, the critical social sciences – anthropology included – evoke (a certain) politics as a strategy to disarticulate (the politics of) science, technique, and everything else of which it is not composed. The problem here is the refusal of the possibility of ontological diversity: for naturalists, rituals of accusation of witchcraft are irrelevant; in the best of circumstances, they are mystifications. The specific drought where such rituals occur is also, in a certain form, irrelevant: what matters are the mathematical, abstract models of reality, in which data that don't fit expected patterns are called *outliers* and discarded. Also for the critical socials scientists, that same drought is irrelevant, as is the specific witchcraft accusation ritual; what matters are the "social processes" in course, that is, sociocultural abstract models of reality. In what has been called *relational ontologies approach*, and whose main inspiration for Brazilian anthropology has been the works of Eduardo Viveiros de Castro (2002), Arturo Escobar (2018), Tim Ingold (2011), and Philippe Descola (2013, 2017), the rejection of the classic dichotomies of modern thought, such as Nature and Culture, Subject and Object, Human and Animal, while transcendent categories, reinserts the material dimension, on one hand, and the irreducible singularity of present contexts, as inescapable elements in the theoretical effort of the social sciences. What is rejected in such an approach is the speculative contemporary metaphysics that thinks the world without humans and humans without the world (Danowski & Viveiros de Castro, 2016).

At the frontier of disasters

I end these reflections by suggesting a research agenda for an Anthropology of Disasters in Brazil, listing themes that seem especially promising. Without losing from sight the justified preoccupation with the performative effect of how events and agendas are framed, I believe it is important to return to the relational ontologies approach, and how they propose the destabilization of the classic Kantian phenomenological references of the constitution of the modern world. What is needed, then, in my view, is the exploration of the consequences of such destabilization. Three dimensions present themselves as fundamental: first, the constitution and status of the *human*; second, of the *subjective*; and third, of the *political*.

In such context, if the human ceases to the mere result of the imposition of arbitrary historical Western distinctions between humanity and animality (Ingold, 2002), and the notion of the subject overflows beyond the species frontiers, it is legitimate to ask this: what are the perspectives of non-human subjects (such as animals) in face of the ontological disorganization of reality brought about by disasters? How can social scientists have access to such perspectives? The methodological challenges here are far from negligible.

Second, the idea that subjectivity, in the ways we are led to constitute and reify it (in Western cultural contexts), is just contingent and not a phenomenological *a priori*, puts in question the need to consider dimensions of existence in which the experience of the world is not mediated by such configurations of subjectivity. As I wrote elsewhere,

> [T]his is one of the most interesting frontiers of the social sciences: from the many variations of what is conventionally called "spirituality" to the phenomenon of crowds, we need to think non-subjectivated and non-subjectivizing ways of being in the world, as a fundamental part of the constitution of the existents (ontologies), without relegating these forms to an "other world."
>
> (Taddei, 2014: p. 604)

And third, the "political" (in its modern, Latourian sense) ceases to be the universal theoretical synchronizer, the great stabilizer of conceptual discourses – similarly to what matter is in the physical sciences. We now live cyborg, asynchronous realities, in which subjects are hybridized with beings of different species, technical objects, computational algorithms, automated processes, and big data: here we witness the displacement of human mediation, and the agent/subject/ego of the social action rarely can be reduced to the individual, in its classic sense. Let's take the Fukushima nuclear accident as an example: the disassembling of reactor number 4 is probably the most dangerous and complex technological task on all human history (Perrow, 2013b), and it can only be done by robots (Perrow, 2013a). In the military field, the use of autonomous drones with lethal power became widespread around the globe. Given the notorious historical ties between spatial and military technologies, and the militarization of the civil defense agencies, particularly in Brazil (Valencio, 2009), the understanding of the (re)composition of worlds and contexts of action is a crucial task for the comprehension of the conditions in which disasters will take place.

One of the consequences of what I propose here is the effort in trying to comprehend the universe of disasters as a form of "post-normal perspectivism of the contemporaneity" (Taddei & Hidalgo, 2016): in altered states – of consciousness, of emotivity, of sensations, of bodily configurations; with fear, anxiety, injury, ontological insecurity – other perspectives impose themselves. The question that arises is, then, this: what other worlds are instituted in such contexts? These states are, obviously, undesired; it does not follow, nevertheless, that they are "abnormal" or "exceptional". Climate change – the tip of iceberg of the Anthropocene – will demand radical recomposition of socionatural (and therefore political) realities (Latour, 2017; Tsing, 2015; Tsing et al., 2017). Under this prism, the need to explore such post-normal perspectives became imperative.

Last, and taking the opportunity to make a reference to how Mexican anthropology engages with the popular cultures of its country,[16] I would like to raise the question of why the world of the dead does not find a place among the legitimate worlds in global debates about disasters and related topics. "Capitalism robbed us everything, even death", a physicist once told me in an ecoliteracy seminar at the Universidad Veracruzana (Veracruzana University), in Xalapa, Mexico – making reference to how Western medical practices disarticulate traditional Mexican ways of "living" death. To take death as a *mode of existence* – something much more possible in Brazil or in Mexico than in France or the United States – presents us with extremely interesting methodological and ontological challenges, not only in our anthropological reflection about disasters; it can be a venue of conceptual development, especially relevant given the challenges predicted for the future of the planet, for which Latin American anthropology may be better suited than other traditions.[17]

Notes

1 The original ideas that eventually became the text here presented profited greatly by discussions held at the Study Group on the Anthropology of Science and Technology (GEACT), at the Federal University of Rio de Janeiro (UFRJ), and at the Research Laboratory on Sociotechnical and Environmental Interactions (LISTA), at the Federal University of São Paulo (UNIFESP). These ideas were developed across more than a decade and a half of ethnographic fieldwork in different regions of Brazil, funded, at different moments, by the São Paulo State Research Foundation (FAPESP, 2007/56394–6, 2007–2009; FAPESP-CLIMAX 2015/50867–8 2016–2020), and the Inter-American Institute for Global Change Research (IAI; CRN3035 and CRN3106, 2012–2017).

2 They are *Mana, Horizontes Antropológicos, Revista Brasileira de Ciências Sociais, Religião & Sociedade, Revista de Estudos Feministas, Vibrant,* and *Etnográfica.*

3 In reality, I am using here a *proxi* for impact factor that is officially used in Brazil: all of the aforementioned journals are at the A1 level (the highest) of the Qualis evaluation system, put in place by the Coordination for the Improvement of Higher Education Personnel (CAPES).

4 I take the opportunity to call the attention of the reader to the fact that the approach adopted in this text leaves absent indigenous views on what is being called disaster, and therefore the account presented here is necessarily incomplete.

5 There is no immediate reading of the joke, as it may enact, depending on the situation in which it is uttered, racism and other forms of discrimination present in popular culture, and/or what Nelson Rodrigues (1997) referred to as a national "stray dog complex" (inferiority complex).

6   Asymmetries in the ways distinctive social groups contribute to the amalgamation of forms of collective imagination of society and the world exist at all levels, including those more markedly "local" (Taddei & Gamboggi, 2009).

7   Freyre and Cascudo were authors born and based in the Northeast region, and whose anthropological and sociological contributions were seminal to the development of the social sciences in the country.

8   Brasília is an exception in such scenario of domination of São Paulo and Rio de Janeiro in the history of anthropology in the country. The presence of anthropology at the national capital is due to the work of two scholars that migrated from the state of Minas Gerais to Rio de Janeiro, and were later involved in the creation of the University of Brasilia: Darcy Ribeiro and Roque de Barros Laraia.

9   These patterns of invisibility existed, naturally, in other, more complex configurations of sociopolitical relations. In the 19th century, during what was perhaps the worst drought ever recorded in Brazilian history (1877–1880), the Ceará state deputy and acclaimed author, José de Alencar, denounced in the chamber of deputies, in Rio de Janeiro, that the accounts of the hecatomb-like drought in his home state, Ceará, that reached the capital of the Empire, could be manipulations by state elites to obtain more funds from the imperial government (Greenfield, 2001).

10 At that time, the idea that there were no hurricanes in Brazil could be found in the pedagogical materials of important public institutions, such as the National Service for Industrial Education (Serviço Social da Indústria-SESI, s.d.); and also in popular science magazines (Revista Superinteressante, 2004).

11 In the case of earthquakes, there are earlier works in the geosciences that document their occurrence in Brazil (see for instance Berrocal et al., 1984), even if outside of the radar of public opinion and of the social sciences.

12 Israel figures prominently in the collective imagination of the inhabitants of the drought prone Brazilian Northeast, as a place where technology has supposedly "won the war" against the environment.

13 With the work of scholars and research groups placed mainly at the Federal University of Minas Gerais (UFMG; see the work of Andréa Luisa Zhouri Laschefski and her colleagues at the GESTA research group; e.g., Zhouri et al., 2016a, 2016b, 2017), Federal University of Espírito Santo (UFES; see the work of Eliana Santos Junqueira Creado and colleagues; e.g., Creado & Helmreich, 2018), Federal University of Ouro Preto (UFOP), and Federal University of Juiz de Fora (UFJF).

14 Sushi and Temaki are widely popular dishes in Brazil, reflecting the existence of a large Japanese-Brazilian community in the country. Most of the salmon consumed in Brazil is farmed in Chile.

15 Mainly due to the work of Norma Felicidade Lopes da Silva Valencio and her collaborators (see, for instance, Valencio, 2004, 2010, 2014; Valencio et al., 2009; Siqueira et al., 2015; Antonio & Valencio, 2016).

16 An earlier and less developed version of this text was first presented at the II Mexican-Brazilian Anthropology Meeting, which took place in Brasília, in 2013. The opening ceremony of the meeting was held at the Mexican Embassy on November 3rd – a day after the Day of the Dead. The participants were received by a diplomat who affirmed having been trained in anthropology. She then conducted everyone to an enormous Day of the Dead altar placed at the entrance hall, and made a detailed and moving description of the elements that constituted the altar, showing that in it pictures of deceased relatives of employees of the embassy, including those of the ambassador and her own, were placed there.

17 An example of that can be found in how Donna Haraway deals with the idea of "dying well" in her 2016 book. Despite of the usual brilliance of the entire argument, she refuses to engage with the concept in the terms of the animistic traditions that she herself mentions in the text, mainly of Native American and Inuit peoples (Haraway, 2016).

References

Albuquerque Jr., D. M. (2014) *The Invention of the Brazilian Northeast*. Durham, Duke University Press.

Antonio, L. S. & Valencio, N. F. L. S. (2016) Animais de estimação em contexto de desastres: desafios de (des) proteção. *Desenvolvimento e Meio Ambiente*. 38, pp. 741–767.

Baniwa, G. (2016) Indígenas antropólogos: entre a ciência e as cosmopolíticas ameríndias. In: Rial, C. & Schwade, E. (orgs.). *Diálogos antropológicos contemporâneos*. Rio de Janeiro, Associação Brasileira de Antropologia.

Beck, U. (1992) *Risk Society: Towards a New Modernity*. Thousand Oaks, CA, Sage Publications Ltd.

Berrocal, J. et al. (1984) *Sismicidade do Brasil*. São Paulo, IAG/USP-CNEN.

Bersani, A. E. (2015) *O (extra) ordinário da ajuda: histórias não contadas sobre desastre e generosidade na Grand'Anse, Haiti*. Master's Thesis, Campinas, Department of Anthropology, State University of Campinas.

Bourdieu, P. (1979) Symbolic Power. *Critique of Anthropology*. 4 (13–14), pp. 77–85.

Candido, D. H. (2012) *Tornados e trombas-d'água no Brasil: modelo de risco e proposta de escala de avaliação de danos*. Doctoral dissertation, Campinas, Geosciences Institute, State University of Campinas.

Catucci, A. (2012) Região de Campinas está na rota de tornados no Brasil, diz Unicamp. *G1*, September 25. Available at http://g1.globo.com/sp/campinas-regiao/noticia/2012/09/regiao-de-campinas-esta-na-rota-de-de-tornados-no-brasil-diz-unicamp.html. Accessed December 14th 2018.

Creado, E. S. J. & Helmreich, S. (2018) A Wave of Mud: The Travel of Toxic Water, from Bento Rodrigues to the Brazilian Atlantic. *Revista do Instituto de Estudos Brasileiros*. 69, pp. 33–51.

Danowski, D. & Viveiros de Castro, E. (2016) *The Ends of the World*. New York, Polity Press.

Deloria Jr., V. (1969) *Custer Died for Your Sins: An Indian Manifesto*. Norman, OK, University of Oklahoma Press.

Descola, P. (2017) ¿Humano, demasiado humano? *Desacatos*. 54, pp. 16–27.

Descola, P. (2013) *Beyond Nature and Culture*. Chicago, The University of Chicago Press.

Durkheim, E. (1982) *The Rules of the Sociological Method*. New York, Free Press.

Escobar, A. (2018) *Designs for the Pluriverse: Radical Interdependence, Autonomy, and the Making of Worlds*. Durham, Duke University Press.

Evans-Pritchard, E. E. (1976) *Witchcraft, Oracles and Magic Among the Azande*. Oxford, Oxford University Press.

Evans-Pritchard, E. E. (1940) *The Nuer*. Oxford, Oxford University Press.

Fonseca, M. K. & Klanovicz, J. (2014) *Comemorar ou esquecer: 25 anos do acidente com o Césio-137 em Goiânia/GO (1987)*. III Simpósio Internacional de História Ambiental e das Migrações, Florianópolis, November.

Gareis, S., da Guía, M., Apolinario do Nascimiento, J., Franco Moreira, A. & Aparecida da Silva, M. (1997) Aspectos históricos de las sequías en el nordeste del Brasil colonial (1530–1822). In: García-Acosta, V. (coord.). *Historia y Desastres en América Latina*. Lima, LA RED-CIESAS-Tercer Mundo Editores, vol. II, pp. 103–132.

Giddens, A. (1991) *Modernity and Self-Identity: Self and Society in the Late Modern Age*. Stanford, Stanford University Press.

Giddens, A. (1990) *The Consequences of Modernity*. Stanford, Stanford University Press.

Greenfield, G. M. (2001) *The Realities of Images: Imperial Brazil and the Great Drought*. Philadelphia, American Philosophical Society.

Haraway, D. (2016) *Staying with the Trouble*. Durham, Duke University Press.

Hoffman, S. M. & Oliver-Smith, A. (eds.). (2002) *Catastrophe & Culture: The Anthropology of Disaster*. Santa Fe, Nuevo México, School of American Research.

Ingold, T. (2011) *Being Alive: Essays on Movement, Knowledge and Description*. New York and London, Routledge.

Ingold, T. (2002) Humanity and Animality. In: *Companion Encyclopedia of Anthropology*. London, Routledge, pp. 48–66.

Kenny, M. L. (2009) Landscapes of Memory: Concentration Camps and Drought in Northeastern Brazil. *Latin American Perspectives*. 36 (5), pp. 21–38.

Kenny, M. L. (2002) Drought, clientalism, fatalism and fear in Northeast Brazil. *Ethics, Place & Environment*. 5 (2), pp. 123–134.

Klanovicz, J. (2010) Apontamentos teórico-metodológicos para uma história ambiental dos desastres "naturais" em Santa Catarina. *Tempos acadêmicos*. 6, pp. 1–18.

Latour, B. (2017) *Facing Gaia: Eight Lectures on the New Climatic Regime*. Malden, Polity Press.

Latour, B. (2013) *An Inquiry into Modes of Existence*. Cambridge, Harvard University Press.

Latour, B. (1991) *We Have Never Been Modern*. Cambridge, Harvard University Press.

Leite, D. A. (2018) *Water Discourses Management: The Communication of the Water Crisis in the Metropolitan Region of São Paulo (2014–2015)*. Master's Thesis, Campinas, Institute of Geosciences, State University of Campinas.

Lopes, A. R. S. (2015) *Desastres socioambientais e memória no sul de Santa Catarina (1974–2004)*. Doctoral dissertation, Santa Catarina, Graduate Program in History, Federal University of Santa Catarina.

Machado, M. H. (2004) The Nature of Tropical Nature: Brazil Through the Eyes of William James. *ReVista – Harvard Review of Latin America*, pp. 13–15, Fall 2004–Winter 2005.

Moreira, P. I. (2013) *Emannuel Liais, Louis Agassiz e Antônio Bezerra: a província do Ceará e os debates geológicos do século XIX*. Paper presented at the XXVII Simpósio Nacional de História, Natal-RN, July.

Nelson, D. R. & Finan, T. J. (2009) Praying for Drought: Persistent Vulnerability and the Politics of Patronage in Ceará, Northeast Brazil. *American Anthropologist*. 111 (3), pp. 302–316.

Oliver-Smith, A. (2010) Haiti and the Historical Construction of Disasters. *NACLA Report on the Americas*. 43 (4), pp. 32–36.

Oliver-Smith, A. & Hoffman, S. M. (1999) *The Angry Earth: Disaster in Anthropological Perspective*. New York and London, Routledge.

Palacios, G. (1996) La agricultura campesina en el nordeste oriental del Brasil y las sequías de finales del siglo XVIII. In: García-Acosta, V. (coord.). *Historia y Desastres en América Latina*. Bogota, LA RED-CIESAS-Tercer Mundo Editores, vol. I, pp. 221–257.

Pennesi, K. (2013) Predictions as Lies in Ceará, Brazil: The Intersection of Two Cultural Models. *Anthropological Quarterly*. 86 (3), pp. 759–789.

Perrow, C. (2013a) Nuclear Denial: From Hiroshima to Fukushima. *Bulletin of Atomic Scientists*. 69 (5), pp. 56–67.

Perrow, C. (2013b) Fukushima Forever. *Worldpost*, September 20. Available from www.huffingtonpost.com/charles-perrow/fukushima-forever_b_3941589.html. Accessed December 14th 2018.

Perrow, C. (1999) *Normal Accidents: Living with High Risk Technologies*. Princeton, Princeton University Press.

Queiroz, M. S. (2017) Dos "tempos do Césio": memória coletiva de um evento crítico. *Resgate: Revista Interdisciplinar de Cultura.* 24 (1), pp. 235–240.

Reis, M. R. C. & Santos, M. E. P. (2017) O desastre em Mariana (MG): expressão da luta pela garantia dos direitos humanos. *Anais do Seminário Científico da FACIG.* 2, pp. 1–7.

Revista Superinteressante. (2004) *Por que o Brasil tem poucos desastres naturais?* April 30. Available from https://super.abril.com.br/ideias/por-que-o-brasil-tem-poucos-desastres-naturais/. Accessed May 3rd 2019.

Rodrigues, N. (1997) *Flor de Obsessão: as 1000 melhores frases de Nelson Rodrigues, organização e seleção de Ruy Castro.* São Paulo, Companhia das Letras.

Rohter, L. (2005) Record Drought Cripples Life Along the Amazon. *The New York Times,* December 11. Available from www.nytimes.com/2005/12/11/world/americas/record-drought-cripples-life-along-the-amazon.html. Accessed December 14th 2018.

Serviço Social da Indústria-SESI. (sd). *Ensino Fundamental – 9a Fase – Tema: Cultura e Ecologia no Século XXI.* Available from http://bit.ly/PYd7QR. Accessed August 4th 2014.

Silva, C. A. M. (2015) Os desastres no Rio de Janeiro: conceitos e dados. *Revista Cadernos do Desenvolvimento Fluminense,* 8, pp. 55–72.

Silva, R. A. C. (2013) *Águas de novembro: estudo antropológico sobre memória e vitimização de grupos sociais citadinos e ação da Defesa Civil na experiência de calamidade pública por desastre ambiental (Blumenau, Brasil).* PhD Dissertation, Rio Grande do Sul, Graduate Program in Social Anthropology, Federal University of Rio Grande do Sul.

Silva, T. C. (2017) Silêncios Da Dor: Enfoque Geracional E Agência No Caso Do Desastre Radioativo De Goiânia, Brasil. *Iberoamericana–Nordic Journal of Latin American and Caribbean Studies.* 46 (1), pp. 17–29.

Siqueira, A., Valencio, N. F. L. S., Siena, M. & Malagoli, M. A. S. (eds.). (2015) *Riscos de desastres relacionados à água: aplicabilidade de bases conceituais das ciências humanas e sociais para a análise de casos concretos.* São Carlos, RiMa Editora.

Strasdas, P. (2011) Respeito é bom e eu gosto. *Blogue Café Puro,* November 6. Available from https://cafepuro.wordpress.com/2011/11/06/respeito-e-bom-e-eu-gosto/. Accessed December 14th 2018.

Taddei, R. (2014) Ser-estar no sertão: capítulos da vida como filosofia visceral. *Interface – Comunicação, Saúde, Educação.* 18 (50), pp. 597–607.

Taddei, R. (2013) Anthropologies of the Future: On the Social Performativity of (Climate) Forecasts. In: *Environmental Anthropology.* New York and London, Routledge, pp. 260–279.

Taddei, R. (2012) Social Participation and the Politics of Climate in Northeast Brazil. In: Latta, A. & Wittman, H. (orgs.). *Environment and Citizenship in Latin America: Natures, Subjects and Struggles.* New York, Berghahn Books, pp. 77–93.

Taddei, R. & Gamboggi, A. L. (2011) Marcas de uma democratização diluída: modernidade, desigualdade e participação na gestão de águas no Ceará. *Revista de Ciências Sociais.* 42 (2), pp. 8–33.

Taddei, R. & Gamboggi, A. L. (2010) Introdução. In: Taddei, R. & Gamboggi, A. L. (orgs.). *Depois que a Chuva Não Veio – Respostas Sociais às Secas na Amazônia, no Nordeste, e no Sul do Brasil.* Rio de Janeiro, Fundação Cearense de Meteorologia e Recursos Hídricos/Instituto Comitas para Estudos Antropológicos.

Taddei, R. & Gamboggi, A. L. (2009) Gender and the Semiotics of Political Visibility in the Brazilian Northeast. *Social Semiotics.* 19 (2), pp. 149–164.

Taddei, R. & Hidalgo, C. (2016) Antropología posnormal. *Cuadernos de Antropología Social.* 43, pp. 21–32.

Thomaz, O. R. (2010) O terremoto no Haiti, o mundo dos brancos e o Lougawou. *Novos estud. – CEBRAP*. 86, pp. 23–39.

Tsing, A. (2015) *The Mushroom at the End of the World*. Princeton, Princeton University Press.

Tsing, A., Swanson, H., Gan, E. & Bubandt, N. (2017) *Arts of Living on a Damaged Planet*. Minneapolis, University of Minnesota Press.

Valencio, N. F. L. S. (2014) Desastres, tecnicismos e sofrimento social. *Ciência & Saúde Coletiva*. 19 (9), pp. 3631–3644.

Valencio, N. F. L. S. (2010) Desastres, ordem social e planejamento em defesa civil: o contexto brasileiro. *Saude soc*. 19 (4), pp. 748–762.

Valencio, N. F. L. S. (2009) O Sistema Nacional de Defesa Civil (SINDEC) diante das mudanças climáticas: desafios e limitações da estrutura e dinâmica institucional. In: Valencio, N., Siena, M., Marchezini, V. & Gonçalves. J. C. (org.). *Sociologia dos Desastres: construção, interfaces e perspectivas no Brasil*. São Carlos, RiMa Editora, vol. 1, pp. 19–33.

Valencio, N. F. L. S. (2004) A produção social do desastre: dimensões territoriais e político-institucionais da vulnerabilidade das cidades brasileiras frente às chuvas. *Teoria & Pesquisa: Revista de Ciência Política*. 1 (44), pp. 67–114.

Valencio, N. F. L. S., Siena, M., Marchezini, V. & Gonçalves, J. C. (orgs.). (2009) *Sociologia dos Desastres*. São Carlos, RiMa Editora.

Vieira, S. A. (2013) Césio-137, um drama recontado. *Estudos Avançados*. 27 (77), pp. 217–236.

Vieira, S. A. (2010) *O drama azul: narrativas sobre o sofrimento das vítimas do evento radiologico do Cesio-137*. Master's Thesis, Campinas, Department of Anthropology, State University of Campinas.

Viveiros de Castro, E. (2002) Perspectivismo e multinaturalismo na América indígena. In: *A Inconstância da Alma Selvagem*. São Paulo, Cosac Naify, pp. 345–399.

Wasserman, R. (1994) *Exotic Nations: Literature and Cultural Identity in the United States and Brazil, 1830–1930*. Ithaca, Cornell University Press.

Weber, M. (1949) *The Methodology of the Social Sciences*. New York, Free Press.

Zhouri, A., Oliveira, R., Zucarelli, M. & Vasconcelos, M. (2017) The Rio Doce Mining Disaster in Brazil: Between Policies of Reparation and the Politics of Affectations. *Vibrant: Virtual Brazilian Anthropology*. 14 (2), pp. 81–101.

Zhouri, A., Valencio, N. F. L. S., Oliveira, R., Zucarelli, M., Laschefski, K. & Santos, A. F. (2016a) O desastre da Samarco e a política das afetações: classificações e ações que produzem o sofrimento social. *Ciência e Cultura*. 68 (3), pp. 36–40.

Zhouri, A., Valencio, N. F. L. S., Teixeira, R., Zucarelli, M., Laschefski, K. & Santos, A. F. (2016b) O desastre de Mariana: colonialidade e sofrimento social. In Zhouri, A., Bolados, P. & Castro, E. (eds.). *Mineração na América do Sul: neoextrativismo e lutas territoriais*. São Paulo, Editora Annablume, pp. 45–65.

3 The Anthropology of Disasters that has yet to be

The case of Central America

Roberto E. Barrios and Carlos Batres

In Memory of Arturo Abilio Berganza Bocaletti,
Irma Yolanda Reyes y Reyes, and Guadalupe Navas de Andrade.

Introduction

Central America, and the nation state of Guatemala in particular, occupy a somewhat paradoxical position in the history of the Anthropology of Disasters. During the early 20th century, disaster researchers and risk reduction experts upheld a view of catastrophes as unavoidable "acts of God" or "nature" that could only be responded to but not prevented (Oliver-Smith, 1999). Beginning in the 1970s, a number of critical social scientists engaged in comparative studies of disaster at a global scale noticed an interesting trend. The magnitude of hazards such as earthquakes and hurricanes did not seem to be the only variable shaping the form (who is affected and how) and severity of disasters. Instead, global data indicated that historical, political, strategic, and socio-economic factors were also critical variables in converting a hazard (e.g., earthquakes, tornados, hurricanes, heavy rains) into a disaster (García-Acosta, 1993, 2002; Hewitt, 1983; O'Keefe et al., 1976; Oliver-Smith, 1999, Oliver-Smith & Hoffman, 1999). An earthquake of magnitude 7 on the Richter scale, for example, could occur in a locality where the absence of socio-economic disparities in access to terrain with low seismic wave magnification qualities and the equitable application of seismic resistant construction techniques prevented societal disruption and the loss of human life and material wealth. On the other hand, when an earthquake of an identical magnitude manifested in the Central American region in 1976, five centuries of inequitable land tenure practices that displaced predominantly indigenous subsistence farmers, imposed poverty among indigenous populations, deforestation, and the introduction of seismic-susceptible urbanization and construction practices produced a disaster that killed 23,000 people and immobilized the nation state of Guatemala. In the same year, a publication by Phil O'Keefe, Ken Westgate, and Ben Wisner in the journal *Nature* made exactly this observation, and this article would become one a handful of scientific publications that founded a new school of thought in disaster studies: vulnerability theory.

Vulnerability theory proposed the idea that a hazard alone did not a disaster make, and that a disaster was the result of a combination of a hazard with a historically and socially created condition of vulnerability that placed certain sectors of a population (differentiated by gender, ethnicity, race, age, and class) at greater risk of harm during and after a catastrophic event. Between the late 1970s and the late 1990s, vulnerability theory became one of the (if not *the*) primary analytical perspectives in the Anthropology of Disasters (Hewitt, 1983; Oliver-Smith & Hoffman, 1999). Two decades after the Guatemalan earthquake, Hurricane Mitch triggered the catastrophic phase of a disaster in Central America. This time, Honduras suffered the lion's share of the damages, although other Central American nation states like Nicaragua, Guatemala, and El Salvador were also significantly affected. As in the case of the 1976 earthquake, Mitch's effects instigated a flurry of research and analysis that enriched the field of disaster studies.

A key tension (or perhaps contradiction) in the field of disaster scholarship, however, is that despite the prominent position of Central America as a fertile region for disaster research, and despite the growth and advancement of disaster anthropology in general, the development of a tradition of disaster anthropology within the region (i.e. the formal training of Central American social scientists in disaster anthropology and their intellectual contributions to the field) remained relatively stunted. In this chapter, we explore the political economic causes of this tension. To accomplish this task, we divide our discussion into three sections. The

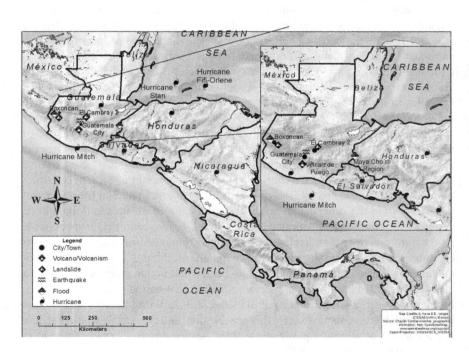

Map 3.1 Map Central America. Case studies and main areas mentioned

first presents a general overview of the development (or relative lack thereof) of anthropology as a discipline in Central America within its broader historical colonial and post-colonial imperialist context. We then move specifically to individually discuss the concerns of disaster anthropology in the two nation states where researchers have conducted the most impactful case studies during the last four decades: primarily Guatemala and, to a lesser extent, Honduras. We note that the limited development of sociocultural anthropology graduate programs in other parts of Central America (i.e. El Salvador, Nicaragua) and, consequently, limited local training of local anthropologists has translated into a relative dearth of disaster anthropology in much of Central America. An attentive reader will notice that the sections this chapter is divided into appear to be out of balance; that is: a significantly longer section is dedicated to the discussion of Guatemala's case than Honduras's, while the rest of Central American anthropology is covered in a single section. We note that this lack of balance is not an oversight on our part, but rather, demonstrative of the inequitable development of disaster anthropology at a global scale. Our seeming lack of balance, then, is representative of academic reality in this part of the Americas.

The development of anthropology in Central America

The limited development of the Anthropology of Disasters in Central America must be understood within the broader context of the history of anthropology in Latin America as a whole. Furthermore, the history of Latin American anthropology cannot be isolated from the political economic and strategic relationships that have shaped life in the continent from the 16th-century era of European colonial expansion to the present. It is noteworthy that Central America remained a sociopolitical littoral area during the Colonial period (D'Ans, 1998; MacLeod, 1973). During the 16th century, Spanish colonizers preferred to settle in areas with high indigenous population densities in order to exploit the latter for their labor and tribute (Carmack et al., 2006; D'Ans, 1998). The result of this settlement pattern was that the centers of colonial society became the regions that are today's Central Mexico in North America and the Andean region in South America. During the pre-Columbian period, a significant portion of the eastern Central American isthmus was home to indigenous populations who maintained much lower population densities than the city state societies of highland Mexico, highland Guatemala, and the Andes. Consequently, colonial society remained relatively underdeveloped in those parts of Central America that did not count among the substantial indigenous populations organized in city-state societies with well-established tribute-extracting social structures. Such was the case of Honduras. In other parts of the Central American isthmus, like highland Guatemala, denser and politically organized indigenous populations did attract higher numbers of Iberian settlers, but this region also remained secondary in sociopolitical importance to central Mexico.

The early colonial period in New Spain is often credited with the emergence of proto-anthropologists; that is, Iberian-born chroniclers and documentarians from the religious orders who set out to document the lifeways and economic systems of indigenous peoples. Among them we can include Bernardino de Sahagún and Diego de Landa. It is worth noting that Sahagún and Landa are just two of a collection of soldier and missionary chroniclers who documented in great detail what they encountered in Central America (see also Fuentes y Guzmán, 1932 [1690]). The displacement of indigenous populations from agriculturally productive lands, imposed conditions of forced labor, high tributes charged by Spanish colonizers, and forced relocations into *reducciones de Indios*[1] brought about the fully anthropogenic disaster of indigenous population decimation during the first half of the 16th century. This catastrophe, which involved epidemics, starvation, and violence, was documented by Bartolomé de las Casas in his *Brevísima relación de la destrucción de las Indias* (2006), which was first published in 1552. These early chroniclers, however, were not formally trained anthropologists and are often not considered part of the canon of the history of anthropology. In fact, historians of the discipline in Central America do not recognize the emergence of Latin American traditions of anthropology until the early 20th century in Mexico and Peru (Bozzoli de Wille, 1994; Pérez de Lara, 1987; Ramírez Cruz & Rodríguez Herrera, 1993).

In the case of Mexico, the history of its anthropological tradition is strongly rooted in the influence of Franz Boas and the eventual Mexican interpretation of his "culturalist" focus by Manuel Gamio. In the Andes, the region's anthropological tradition was heavily influenced by German diffusionism and Alfred Kroeber's version of this same theoretical school. Guatemala, on the other hand, did not participate in this early 20th-century trend and remained without an anthropological tradition until the "revolutionary period"[2] of 1944–1954 with the establishment of the *Instituto Indigenista Nacional* (National Indigenist Institute) in 1945, the *Instituto de Antropología e Historia de Guatemala* (Institute of Anthropology and History of Guatemala) in 1947, and the creation of the geography and history tracks within the Department of Humanities at the *Universidad de San Carlos de Guatemala* (University of San Carlos of Guatemala), also in 1945. These three institutions were created as part of a broader strategy by the revolutionary government to a) document the historical and social conditions of Guatemala, and b) to conduct studies of indigenous populations for the establishment of a national history (Pérez de Lara, 1987).

The ousting of the democratically elected president, Jacobo Arbenz, by a US-backed right wing military coup under the direction of Carlos Castillo Armas sidetracked the role anthropology was to play in nation building under the revolutionary government. What followed the 1954 coup was the continuation of what some (but not all) Guatemalan anthropologists call *antropología de la ocupación* (anthropology of occupation), which had also been present in the country in the era preceding the 1944 revolution (Pérez de Lara, 1987). Anthropology of occupation refers to research and scholarship conducted by US-born and trained anthropologists who focused on community studies and promoted a colonialist agenda focused on the integration of indigenous populations into the national society.

Whether the term *antropología de la ocupación* is a fair label to use when describing US-born, trained, and based anthropologists during the period preceding and directly following the revolution of 1944 remains a topic of significant debate among prominent Guatemalan anthropologists. Olga Pérez de Lara (1987), for example, uses the term to refer to the work of Richard Adams, claiming he subscribed to a structural-functionalist approach that ignored political economic issues prior to the publication of his celebrated text, *Crucifixion by Power: Essays on Guatemalan National Structure, 1944–1966* (1970). Meanwhile, other scholars like Roberto Melville (personal communication) argue that the reading of Adams' pre-1970 work as structural functionalist is an erroneous oversimplification, and that the term *antropología de la ocupación* is sometimes used as a nationalist response on the part of some hardline Guatemalan Marxist anthropologists against all scholarship originating from United States institutions (see also Bolaños Arquin, 1994). Determining whether use of *antropología de la ocupación* is warranted or not remains out of the scope of this chapter, but it is worth noting that key volumes in the anthropology of Guatemala like Sol Tax's *Penny Capitalism* (1953), whether "occupationist" or not, subjected Mesoamerican indigenous communities to a North American ethnocentric reading that painted their lifeways and cultural practices in a light that celebrated capitalism as a human universal.

The development of Guatemala's anthropological tradition, however, was interrupted by, of all things, the catastrophic phase of a disaster triggered by an earthquake in 1976, which killed tens of thousands of people and disproportionately affected rural indigenous communities. The catastrophe coincided with a growing desire within Universidad de San Carlos (San Carlos University) to consolidate an undergraduate degree program in sociocultural anthropology that included a fieldwork component. The first generation of anthropologists in this program consisted of five students who, when confronted with the conditions of life among marginalized indigenous communities, became committed to the application of anthropology in addressing Guatemala's stark socioeconomic inequities. The response of the Guatemalan state to this grassroots development of applied anthropology was a violent reactionary one, resulting in the murder of three of the five students (Arturo Abilio Berganza Bocaletti, Irma Yolanda Reyes y Reyes, and Guadalupe Navas de Andrade) by security forces and the self-imposed exile of the remaining two (Pérez de Lara, 1987).

Disaster and the development of a Guatemalan tradition of anthropology, then, are intimately intertwined. Unfortunately, the violent repression of academics by the US-backed military regimes from the late 1970s to the 1990s stunted the development of anthropology in Guatemala. Furthermore, the profound shock of the Guatemalan state's repression of progressive civil society, indigenous communities, and insurgent movements demanded the attention of what little anthropological tradition did develop, keeping other anthropological concerns relatively underdeveloped. The violent murder of anthropologist Myrna Mack in 1990 for her work among internally displaced indigenous populations reaffirmed the pattern of state repression established in the late 1970s. Despite these repressed beginnings, recent undergraduate theses from Universidad del Valle (Del Valle

University) and edited volumes published by Universidad de San Carlos have revived a concern with disaster among Guatemalan anthropologists. As we will demonstrate in the section that follows, Guatemalan disaster anthropologists recognize the historical political ecological roots of disasters, but much of their work primarily focuses on civil conflict and haphazard urbanization, which are nevertheless inevitably tied to disasters in the region.

As we noted earlier, the settlement of New Spain during the colonial period primarily followed the contours of state-level indigenous societies. In the region that is today's Central America, 16th-century indigenous state societies concentrated mostly in the Guatemalan highlands and western Honduras and El Salvador. In other parts of Central America, indigenous populations were more loosely politically organized and featured lower population densities. In the absence of densely concentrated indigenous people to exploit, Spanish settlers avoided much of eastern Central America in their settlement patterns, leaving broad parts of Honduras, El Salvador, and Nicaragua at the margins of colonial society (D'Ans, 1998; MacLeod, 1973). This colonization pattern had long lasting effects that are still seen today in the sociopolitical conditions of these nation states as well as the development of national anthropological traditions. In 1993, Ana Lilian Ramírez Cruz and América Rodríguez Herrera described the state of the development of the discipline of anthropology in El Salvador thus: "In El Salvador we cannot speak beyond a precarious and incipient development of anthropology due to conditions in the social and political order, which have determined severely reduced spaces for the development of anthropology" (Ramírez Cruz & Rodríguez Herrera, 1993: p. 37).

Just as in the case of Guatemala, the period of Cold War-related state violence in the 1980s negatively affected the development of anthropology in Honduras and El Salvador. During this decade, Honduras experienced US-backed state repression of critical academic voices and members of civil society. Despite these challenges, there have been some advances in the development of a Central American anthropological tradition, albeit limited ones. The 1950s and '60s saw the establishment of an academic university program focused on social sciences thanks to the efforts of Alejandro Dagoberto Marroquín, a program that focused specifically on the improvement of living conditions among indigenous communities in El Salvador. In Honduras, Mexican anthropologist Gonzalo Aguirre Beltrán promoted applied anthropology through a seminar at the National Autonomous University of Honduras in 1968. The foci of these efforts were technical assistance, health, and development in rural communities (Bozzoli de Wille, 1994). Finally, in Costa Rica, the mid-1970s witnessed a growing interest in the application of anthropology as a means of social action. Disasters, then, remained a peripheral, if not an altogether invisible, topic in Central American anthropological research.

The Anthropology of Disasters in Guatemala

The key hazards that affect the Guatemalan population are floods, earthquakes, hurricanes, volcanic eruptions, and landslides. However, to understand how these hazards are transformed into the catastrophic phases of disasters, we must employ a historical political ecological lens that lets us see the relationship between colonization, land use, urbanization, and "development" practices that have taken place in the region for the last 500 years.

Between 1544 and 1560, the Spanish crown instituted the *reducciones de Indios* that required indigenous people to move from their traditional communities into localized settlements, guaranteeing early Spanish *conquistadores* a steady supply of taxes and solidifying a workforce for use on their *haciendas*. These policies essentially segregated the indigenous population from the early Spanish social sphere, and subsequently from the colonial sociopolitical and economic spheres (Martínez Peláez, 2009). Thus, indigenous people were kept separate and away from Spanish settlements, relegated to areas the Spanish considered rural. As Spanish architects designed new colonial spaces, they introduced European architectural techniques and aesthetics, material construction, and city planning, helping them reinforce the "ruralization" of indigenous communities and relegating them to haphazard living conditions (Guzmán Böckler & Herbert, 1970; Martínez Peláez, 2009). Pre-Columbian rural indigenous settlement patterns followed a scattered village pattern, which allowed ample separation between households. The *reducciones* imposed squalor on many indigenous families and created denser settlements that, when combined with imposed malnutrition, forced labor, and new pathogens like smallpox, resulted in virulent epidemics that destroyed large parts of the native population. Thus, the ways that the rural and the urban were settled early on introduced vulnerabilities for Mayan communities that would continue throughout the colonial era.

Following independence in 1821, and then a governmental overthrow in 1871, *ladino* (non-indigenous) political leaders instituted new policies known as the *Reforma Liberal* (Liberal Reforms). The economic objective of these reforms was to incorporate the young nation state of Guatemala into the international capitalist system of commercial trade by establishing coffee as the main crop cultivated for export (Wagner et al., 2001). These reforms guaranteed the production of coffee for the following eight decades by enabling the appropriation and consolidation of lands acquired from indigenous communal holdings and outlining new regulations regarding indigenous workers. This resulted in the indigenous population being subjected to a forced displacement from their communities and facilitated their exploitation as a workforce for the coffee plantations (Alvarez, 1994). Thus, the *Reforma Liberal* redistributed land in new ways, creating a bifurcated system of landholdings (large *latifundio* estates versus small *minifundio* subsistence plots) and increasing the indigenous population's colonial inheritance of social and economic vulnerabilities. This unequal distribution of land and the continuing exploitation of the indigenous contributed heavily, along with many other social and economic factors, to the start of the civil war in Guatemala in 1960. This war, it is worth noting, featured a clandestinely US-backed ethnocidal counter-insurgency campaign that specifically targeted rural Maya communities. During this conflict, which would last for the next 36 years, it was the already vulnerable indigenous population that would pay the greatest price in terms of lives lost (over

166,000 of the 200,000 killed), disappearances, and displacements (Guatemala, Memory of Silence, 1999).

In this historical context, and amidst the backdrop of the ongoing civil war, an earthquake of 7.5 Richter scale magnitude shook Guatemala in 1976, resulting in the loss of thousands of human lives (23,000) and almost half of the country's infrastructure. These losses could not be reduced to the direct impacts of a seismic hazard alone. They were the result of changes made in the region's construction materials and styles by early Spanish architects (Bates et al., 1979), the consolidation of the Spanish urban and the indigenous rural dichotomy in colonial times, and the ongoing historic disenfranchisement exacerbated by the *Reforma Liberal*, in which official appropriation and redistribution of agrarian resources limited land access for and displaced the indigenous population.

The 1976 earthquake did not only strike an already vulnerable population, its aftermath also enhanced existing inequalities and vulnerabilities and created new ones in the form of systematic violence as the country's civil war peaked in the late 1970s and early 1980s. This violence was enacted with two key objectives. The first was stopping guerrilla-backed social movements that favored increased indigenous rights and unionized labor, which were seen as threatening the country's existing economic and social order (Aguilar, 2006). The second was to redraw the social landscape of the Guatemalan countryside by "modernizing" indigenous communities and bringing them within closer control of the Guatemalan state. A key element of this second agenda was the creation of development poles that, like the *reducciones* of the colonial period, attempted to condense indigenous settlements in a manner that facilitated population management for state institutions. The civil war and the 1976 earthquake shed light on the relationships between historic, social, political, and economic vulnerabilities and hazards that give disasters form and magnitude. The convergence of these processes and events can serve as a frame of reference for understanding how population displacement, ethnic and cultural segregation, lack of urban planning development, and lack of access to living spaces intersect with earthquakes, hurricanes, floods, and volcanic eruptions to produce the catastrophic phases of disasters.

Foreign aid in the years following the 1976 earthquake, combined with a domestic economic upturn, promoted the growth of Guatemala City and added to its increasing concentration of job opportunities immediately after the disaster. However, the extensive recovery processes, resettlement attempts, and reconstruction were only a relative success (Bates et al., 1979; Bates, 1982). The concentration of recovery efforts in the city, coupled with the disastrous effects of the ongoing civil war's peak in the rural areas in the late 1970s and early 1980s, compelled many people from rural areas and smaller towns to move into Guatemala City. However, plans for the city's reconstruction did not take into consideration that the workforce required for the recovery and simultaneous urban migration would need adequate living spaces exposed to minimal hazards. A lack of available housing (and eventually a lack of opportunities after the city's population exploded and reconstruction efforts slowed) led to the growth of unplanned settlements, which cropped up in whatever marginal spaces were available.

With much of Guatemala City in ruins, many members of the recovery work-force and post-earthquake migrants to the city were forced to establish their homes in areas that were not suitable for habitation, such as the *barrancos*, or ravines, that cut across this urban area. These precarious settlements would become known as *Asentamientos Humanos* (irregular human settlements) (Bastos & Camus, 1994, 1995). Today, these parts of the city of Guatemala have become permanent urban neighborhoods located in marginal places that are settled by a population comprised of many diverse ethnic and cultural groups including indig-enous Maya, *Garifunas* (Afro-Caribbeans), and *Ladinos* (individuals considered non-indigenous) who share a common displacement history and limited access to economic resources (Miner, 2002).

The *Asentamientos Humanos* are commonly portrayed, often unfairly, as mar-ginal neighborhoods of urban survival populated by delinquents and miscreants. In addition, these places have been politically characterized as resistant to change and their population interpreted by authorities as having a pathological opposition to the benefits of development (Miner, 2002). These distillations and stereotypes erase or obscure the historical political economic processes though which *Asien-tamentos Humanos* developed; namely, ethnic discrimination, civil war displace-ments, and urban planners' lack of concern or the social needs of the migrants and workers who moved from rural areas to Guatemala City in the years after the earthquake. Recently, the work of Guatemalan anthropologists (Aguilar, 2006) has demonstrated how this population, forced to inhabit marginal and hazardous spaces, became even more vulnerable not just to earthquakes, but also to other natural hazards or extreme events common to Guatemalan physiography, such as volcanic eruptions and hurricanes.

In 1998, the devastation caused by Hurricane Mitch was so widespread through-out Central America that it created a new way to visualize and label disaster as a "regional disaster." In this regard, not only was the disaster seen on televised radar to encompass the entire isthmus, but was quantified in terms of economic losses that surpassed 7 billion dollars, and more than 20,000 lives were believed to have been lost (Donado, 2008). One of the populations most affected by Hur-ricane Mitch in Guatemala was the Maya Cho'rti' community on the eastern side of the country and whose settlement extends into Honduras. Over 2,500 Cho'rti' were killed by mudslides triggered by the rains associated with Hurricane Mitch (Marshall, 2007).

Though it was not as strong as Mitch, the severity of rain caused in 2005 by Hurricane Stan also caused a loss of infrastructure and extensive mudslides that exacted a large cost in terms of human lives. In Guatemala, the number of deaths and disappearances from the hurricane itself was close to 2,000, but resulting flood-ing and mudslides also killed many in the highlands of Guatemala. In fact, the mud-slides caused by Stan were so severe that the entire Maya community of Boxoncán was buried. The destruction was so complete that the bodies of the inhabitants sim-ply could not be extracted, and it is now classified as a graveyard (Aguilar, 2006).

Maya communities in Guatemala are more vulnerable to mudslides and flood-ing caused by hurricanes because this population has been denied access to their ancestral lands through historical displacement and racial disenfranchisement. They have been forced to inhabit terrain at the highest elevations of mountainous areas with little agricultural productivity in contrast to their ancestral, productive land (in the valleys and piedmonts) that is now privately owned and dedicated to coffee farming and cattle ranching (Garrido, 2007). The research of Guatemalan anthropologists (Aguilar, 2006; Garrido, 2007) has illustrated how hurricanes Mitch and Stan put on display, once again, the historical social inequities related to indigenous communities and the lack of governmental attention to these communities' vulnerabilities when they are struck by disaster. Furthermore, they have reiterated these communities' lack of access to disaster hazard-free living spaces (Aguilar, 2006; Garrido, 2007).

Another example of how a lack of access to safe living spaces (as well as negligence and mismanagement of disaster resources) can make communities more vulnerable is the case of the El Cambray 2 community. A large part of El Cambray 2, which used to lie south of Guatemala City, was buried by a mudslide in 2015 when a nearby hill collapsed, killing over 300 residents at once. An additional 70 people disappeared, and dozens were left homeless. One survivor described how "sand began to fall on the metal roof upstairs and [he] began to smell dirt in the air and heard a buzzing noise like when the waves crash into the shore. Everything was demolished" (Patzán, 2017). One of the first residents to return home after disaster struck described arriving at the scene to find a large group of people assembled near his home. He explained that at first he

> thought it was a fight, but the more I walked the more people I saw, I got nervous and borrowed a cellphone to call all three of my family members, but they kept going to voicemail. Then I knew what happened and began to panic. I grabbed a shovel and I dug in the place where I knew the house was. I dug for three meters, but at three in the morning my arms gave out. I lost my wife and three children.
>
> (Patzán, 2017)

The first houses built in El Cambray 2 were constructed in 1994, and a government evaluation carried out in 2014, one year before the disaster, warned that the community was at risk of possible landslides (Castañaza, 2015). However, because relocating the people in these areas would equate to lost tax revenue for the local municipality, local officials chose to ignore the warnings. After the landslide, state authorities indicted two local mayors for not initiating actions to move the residents to a safer place. They were both accused of ignoring critical warnings indicating the high degree of danger to which people living in this area were exposed (Orozco & Ramos, 2016; Ramos, 2016). Disasters such as the mudslide at Cambray 2 have been linked to unplanned development processes including urban sprawl, poor planning, and limited access to hazard-free spaces for housing. These problems are further compounded by negligence in disaster preparedness, including officials not following codes or heeding warnings, as well as the all-too-common mismanagement of post-disaster aid finances.

Issues surrounding a lack of disaster preparedness and ineffective disaster recovery plans have reemerged in the tragedy that resulted from a recent volcanic eruption. On the morning of June 3, 2018, the inhabitants of the communities near the Volcán de Fuego awoke to hear rumbling, but accustomed to the active volcano they lived beside and hearing no official warning, they continued with their normal daily activities. The behavior of the volcano seemed ordinary, and no one imagined that a few hours later a life-changing eruption would bring calamity. The eruption around noon was so strong that the flows of pyroplastic matter, mud, debris, and water reached 600 meters in length. Victims were trapped in their homes when they were hit by this dangerous flood of lahar, burying them and their surrounding communities, and those fortunate enough to have survived the onslaught of the eruption escaped with third-degree burns.

The eruption of the Volcán de Fuego became another disaster of great destruction and pain for Guatemalans, with the official total declaring 113 dead and 329 missing. It is believed that nearly 2,900 people were killed and, in all, over a million people in Guatemala were affected by the eruption, which resulted in economic losses in excess of 200 million dollars (Paredes & Morales, 2018). Following the eruption, lawmakers accused CONRED (the National Coordination for Disaster Reduction) and INSIVUMEH (the National Institute for Seismology, Vulcanology, Meteorology, and Hydrology) of potential criminal negligence for mismanagement and insufficient warnings in advance of the disaster. Despite knowledge of the eruption eight hours before it occurred, official warnings for mandatory evacuation were issued far too late (three hours after the eruption).

In the days that followed, the United Nations Economic Commission for Latin America and the Caribbean warned that that the survivors of this disaster were now at risk for becoming further impoverished due to the disruption of their way of life and displacement from their homes (Gamarro, 2018). Yet despite an initial influx of national and international aid for assisting the victims of this disaster, only months afterwards this risk became even more dire and recovery was incomplete. As columnist Samuel Reyes Gómez (2018) described, "the victims of the eruption of the Volcán de Fuego have already been forgotten, despite the fact that they continue to suffer the consequences and their extreme levels of poverty continue to grow".

Guatemalan anthropologists have also explored the meanings their compatriots give to historical processes of spatial, economic, ethnic, and political exclusion, especially when these old wounds of exclusion are reopened by the occurrence of earthquakes, mudslides, hurricanes, and volcanic eruptions. Although disaster survivors continue to reaffirm their understandings of their historically and political-economically imposed vulnerabilities, many also attribute catastrophic occurrences to acts of God or divine intervention (Miner, 2002). Divine explanations may be a way of coping with feelings of powerlessness in the face of both political-economic impositions and severe hazards. For members of vulnerable populations, rather than worrying about how to avoid historical political ecological processes they alone cannot control, worrying about their more immediate, daily survival needs (i.e., how to procure better housing and constant barriers to education, health care, and job opportunities) becomes a reasonable strategy for pre- and post-disaster planning, or a lack thereof. Others in vulnerable communities have combined their ancestral cultural knowledge with their knowledge of the modern world in explanations of their collective inability to avoid a catastrophic event that devastated their community. Consider the words of the words of this Mam Maya interlocutor in the aftermath of Hurricane Stan:

> The people say that it's a big animal whose name is Waxkan [a monster that is half serpent and half bull], which opened the land and exploded the earth . . . but I don't know, it could be that great men opened a hole in the land and the sea, which extracted the water and the wind that was produced came towards us . . . with equipment, with tools and helicopters, they opened the hole in the sea.
>
> (Aguilar, 2006: p. 44)

These existing senses of vulnerability to disastrous events are exacerbated when survivors are not even able or allowed to recover and provide an adequate burial for their deceased loved ones according to their own cultural and religious practices (Donado, 2008). Such is the case of complete communities being declared cemeteries. These instances deny survivors the social and cultural rites necessary for emotional closure and impede the therapeutic processes survivors must go through in order to begin healing, rebuilding their community, and reinforcing a sense of control over their surrounding environment (Holland & van Arsdale, 1989). Survivors' storytelling implies, however, that it is the sense (but not the actual state) of powerlessness in the face of disasters that engenders even greater vulnerability and puts them at the mercy of shifting political agendas and in the hands those who manage plans for recovery that are often only half-completed. Thus, survivors are caught in the clash between historical exclusions and physiographic phenomena, and attention to their needs and vulnerabilities seem to reappear and disappear along with the media coverage of the earthquakes, mudslides, hurricanes, and volcanic eruptions that they experience. It is in the storytelling about these shifting political momentums and how plans for recovery are managed that Guatemalan anthropologists have found new meanings given to disasters.

Guatemalan anthropologists now consider disasters to be the result of interactions between historical political ecological processes and geophysical events, whose materialization (as an earthquake, mudslide, hurricane, or volcanic eruption) takes place at a particular location and within a specific temporal and cultural context (see also García-Acosta, 2018). Thus, it has become important for anthropologists in Guatemala to examine the past history and present reality of Guatemalan communities' vulnerability to disasters (Aguilar, 2006; Donado, 2008; Garrido, 2007; Miner, 2002). These same anthropologists have also discussed how, during post-disaster periods, affected communities have become politicized contexts where multiple individuals and social institutions compete to manage projects intended to assist in moments of crisis or prevention of future disasters.

The communities and populations which are intended to receive help are typically framed according to reconstruction criteria that were often designed by outsiders in a different cultural context. In these scenarios, reconstruction 'assistance' is thus delivered in line with particular political economic agendas. Therefore, humanitarian aid becomes encased a structure of power based on political conditions, as well as a host of internal and international strategic relationships that form an invisible appendage of the modern (neoliberal) economic system (Aguilar, 2006).

The Anthropology of Disasters in Honduras

While some scholars often interpret the 1976 Guatemalan earthquake as an intensifying event that further complicated civil unrest and may have very well detonated the most violent phase of the country's civil war (Way, 2012), disasters do not seem to feature so prominently in the history of Central American anthropology elsewhere in the isthmus. It is worth noting that some parts of Central America were heavily affected by disasters during this time. For example, Hurricane Fifi in 1974 severely impacted Honduras, but what social science research exists was conducted by US-based sociologists and focused mostly on aid delivery and assessment rather than a more critical understanding of the root causes of disaster and vulnerability (Snarr & Brown, 1980).

Throughout the 1980s and 1990s, Honduras would become one of several key sites in the development of political ecological analytical perspective that, in turn, played an important role in the popularization of vulnerability theory in US disaster anthropology circles. In this case, it is not Central American anthropologists who are the protagonists of the story, but North American and European researchers (Jansen, 1998; Paolisso et al., 1999; Stonich, 1993). Much of this work would provide a morbid foreshadowing of what was to come in October 1998 when Hurricane Mitch coincided with 500 years of inequitable land tenure practices, poorly planned urbanization, and deforestation to produce one of the most devastating catastrophes of the 20th century. Honduras was the hardest hit of the Central American nations, reporting a loss of 70% of its GDP, the destruction of 35,000 homes, and the deaths of at least 6,000 people (PAHO, 1998).

Hurricane Mitch's timing was particularly interesting in that, since the last Hurricane-related disaster in Honduras (Hurricane Fifi), disaster anthropology had come of age in various American academies, including Mexico and the United States. While Honduras itself may not have counted with a robust cadre of disaster researchers, it did become a site where a variety of US and Canadian-trained anthropologists (and other related social scientists) set out to apply anthropological theories and methods in the hopes of enhancing both recovery efforts and the state of the art in anthropological knowledge about disaster recovery. Among these social scientists were transnational Central American researchers like Roberto E. Barrios and Vilma Fuentes who worked in collaboration with Anthony Oliver-Smith and James P. Stansbury from the University of Florida (Ensor, 2009b). Anthropological (and related social science) research conducted on post-Mitch reconstruction in Honduras, El Salvador, and Nicaragua has focused heavily on community relocation and housing reconstruction projects (Barrios, 2009a), the response of civil society to the disaster (Fuentes, 2009), gender (Bradshaw, 2001; Cupples, 2007; Ensor, 2009a), and NGO-local government-disaster survivor power relations (Barrios, 2014, 2017a).

Anthropological studies of Mitch in Honduras revitalized the Anthropology of Disasters in the United States. Previous anthropological research on community relocations (Cernea, 1996) had rightly highlighted the socioeconomic impacts of such programs on affected populations but had failed to recognize the relationships between culture and power in governmental and aid agency policy and practice (i.e., who gets to imagine policy and practice, what imagination makes policy and practice possible, and what is the cultural history of such an imagination). Disaster reconstruction and community relocation projects in Honduras provided ethnographic fieldwork experiences for a new generation of disaster researchers that forced them to confront the inherent assumptions of disaster reconstruction policy and the often deleterious impacts of such assumptions on the lifeways of disaster survivors. Furthermore, disaster anthropology had remained dependent on political ecology as its primary analytical perspective, but the case of Honduras beckoned disaster anthropologists to come into conversation with poststructural critiques of development, the anthropology of space and place, and affect theory (Barrios, 2017a). Ethnographies of the recovery process illustrated how disaster recovery features a number of encounters between social actors (disaster survivors and their grassroots leaders, NGO staff and program managers, urban planners, local and national government officials) each with their own practices and ideas of successful reconstruction. The anthropological analysis of such encounters required disaster anthropologists to consider critiques of development discourse articulated by the likes of Arturo Escobar (1995), James Ferguson (1994), and James Holston (1989) during the late 1980s and 1990s. Additionally, the natural discourse of disaster survivors in post-Mitch Honduras routinely mobilized a language of affect and emotions to assess the efficacy of recovery programs, demanding a combination of disaster anthropology with affect theory. By requiring researchers to engage more openly with other current trends in anthropology, the anthropology of Mitch has helped highlight the role of disaster anthropology as a field of theoretical innovation (see also Barrios, 2017b).

To our knowledge, however, Hurricane Mitch did not have the same effect the 1976 Guatemalan earthquake had in stimulating the development of Central American disaster anthropology. The reasons for this differential effect are beyond our ability to determine at the moment with the existing literature, and identifying them would certainly prove to be a worthwhile topic of investigation for scholars interested in the history of the field in this region. Despite our limited knowledge, we would like to take the liberty to hypothesize that Guatemala in 1976 was a very different sociopolitical context than Honduras in 1998. The 22 years that separated the disasters saw significant US-supported state terror that decimated civil society leaders and critically engaged scholars throughout the region. The result was an absence of local scholars and a lack of economic support for local scholarship that could have taken on the task of doing the anthropology of Mitch.

Final reflections: the extractive economy of social science research in Central America

Since the late 19th century, Central America has provided US and European anthropologists with seemingly limitless opportunities for professional development in the form of field sites, case studies, and the hospitality local interlocutors and fellow academics have shown them. Central America has also provided data for research that has greatly advanced anthropological knowledge in general and the Anthropology of Disasters, as well. Unfortunately, the creation of anthropological knowledge in the region remains an extractive economy where scholars and researchers from hegemonic nation states develop careers and publications thanks to the human and ethnographic resources of Central America, while the development of local anthropology programs and traditions lags dismally behind. In the case of Guatemala, research on the effects of colonial, structural, and state violence became the primary preoccupation of Guatemalan-born anthropologists during the late 1970s and 1980s, costing several of them their lives, while US-born anthropologists who knew about the violence simply suspended their fieldwork seasons and avoided the topic altogether until much later. The imperial privilege of US anthropologists on the topics they chose to research and the ability to spatially distance themselves from state violence were not shared by their Guatemalan and Central American counterparts, and the decimation of early generations of trained anthropologists and the message of terror their assassinations sent future generations has had a long-lasting impact on the development of the field in the region.

As we write this chapter in December 2018, we are witnessing the long-term effects of US-backed state violence in the isthmus. The persecution of civil society and labor organizers in Guatemala, Honduras, and El Salvador, and the US-supported counterrevolutionary movement (i.e., the Contras) in Nicaragua effectively crippled the ability of Central American civil society to organize itself in a way that allowed possibilities for more equitable societies and avenues for upward social mobility for historically disenfranchised groups (Barrios, 2009b; Way, 2012). In addition, the influx of decommissioned former soldiers and revolutionary fighters into societies that were not equipped to provide them with the economic and psychological support they needed made an already difficult situation even worst. Without the resources necessary to help them make a transition to civilian life, decommissioned forces have contributed to the problems of escalating social insecurity and violence. The result has been the emergence of societies that are increasingly inhospitable to their own residents, forcing them to choose migration to the United States in search of economic stability and security. While migration from Central America and Mexico is increasingly defined as a "problem" that assails the United States from without, it is imperative that the US public and law makers recognize a) the key role of US covert operations and foreign policy in the creation of this "problem" and b) the ethical and moral responsibility of the US to look after the social needs of populations whose societies it disrupted.

Just as the immigration "crisis" in the US cannot be understood without knowledge of the history of US clandestine practices and political and military influence in Central America, the stunted history of anthropology and disaster anthropology in the region must also be framed within the context of strategic imperial relationships in this part of the hemisphere. Consequently, we feel that the US State Department and US academic institutions also have a moral and ethical responsibility to financially and institutionally support the development of Central American anthropology well beyond current levels. Transnational Central American anthropologists like Roberto Melville of the Centro de Investigaciones y Estudios Superiores en Antropología Social (Center for Research and Advanced Studies in Social Anthropology, CIESAS) in Mexico City have worked tirelessly for the development of postgraduate programs in anthropology in Central America, only to find their efforts systematically underfunded and undersupported by US public and private institutions. We therefore make a call for agencies like the Wenner Gren Foundation and the National Science Foundation to launch initiatives that do not focus primarily on the development of American anthropologists, but that engage in what we would like to call "academic justice," that is, the remediation of the effects of colonial and capitalist imperial violence in regions like Central America.

Acknowledgments

The authors thank Roberto Melville for his review of this manuscript and his helpful critiques and recommendations.

Notes

1 *Reducciones de Indios* were settlements populated by forcefully relocated indigenous populations. Their purpose was to make the labor and tribute of their residents readily available to Spanish colonizers (Carmack et al., 2006; MacLeod, 1973). The *reducciones* also facilitated the political and military control of their residents.
2 Which was a response to the excesses of dictator Jorge Ubico's presidency during the preceding 13 years (Way, 2012).

References

Adams, R. N. (1970) *Crucifixion by Power: Essays on Guatemalan National Social Structure, 1944–1966*. Austin, University of Texas Press.
Aguilar, S. J. (2006) *Anatomía Social de un Desastre: Consecuencias Sociales Asociadas al Impacto de la Tormenta Stan en Una Comunidad del Altiplano Occidental de Guatemala*. BA Thesis. Guatemala, Universidad del Valle de Guatemala.
Alvarez, O. (1994) Antecedentes Históricos del Proceso Revolucionario de 1944–1954 en Guatemala. In: Velásquez, A. (ed.). *La Revolución de Octubre: Diez Años de Lucha por la Democracia en Guatemala, 1944–1954*. Guatemala, Centro de Estudios Urbanos y Regionales (CEUR), Universidad de San Carlos de Guatemala.

Barrios, R. E. (2017a) *Governing Affect: Neoliberalism and Disaster Reconstruction.* Lincoln, University of Nebraska Press.

Barrios, R. E. (2017b) What Does Catastrophe Reveal for Whom? The Anthropology of Crises and Disasters at the Onset of the Anthropocene. *Annual Review of Anthropology.* 46, pp. 151–166.

Barrios, R. E. (2014) Here, I'm Not at Ease: Anthropological Perspectives on Community Resilience. *Disaster.* 38 (2), pp. 329–350.

Barrios, R. E. (2009a) Tin Roofs, Cinder Blocks, and the Salvatrucha Streetgang: The Semiotic-Material Production of Crisis in Post-Hurricane Mitch Reconstruction. In: Ensor, M. O. (ed.). *The Legacy of Hurricane Mitch.* Tucson, University of Arizona Press, pp. 157–183.

Barrios, R. E. (2009b) Malditos: Streetgang Subversions of National Body Politics in Central America. *Identities.* 16 (2), pp. 179–201.

Bastos, S. & Camus, M. (1995) *A la Orilla de la Ciudad.* Guatemala, FLACSO.

Bastos, S. & Camus, M. (1994) *Sombras de una batalla. Los desplazados por la violencia en la Ciudad de Guatemala.* Guatemala, FLACSO.

Bates, F. (1982) *Recovery, Change and Development: A Longitudinal Study of the 1976 Guatemalan Earthquake.* Athens, GA, The University of Georgia Press.

Bates, F, Farrell, T. & Glittenberg, J. K. (1979) Some Changes in Housing Characteristics in Guatemala Following the February 1976 Earthquake and Their Implications for Future Earthquake Vulnerability. *Mass Emergencies.* 4, pp. 121–133.

Bolaños Arquin, M. (1994) *Anthropological Approaches in U.S. Studies of Central America, 1930–1970: Implications for Central American Anthropology.* Doctoral Dissertation. Lawrence, University of Kansas.

Bozzoli de Wille, M. E. (1994) La Antropología Aplicada en Costa Rica y en Centroamérica. *Revista Reflexiones.* 22 (1). Available at https://revistas.ucr.ac.cr/index.php/reflex iones/article/view/10734. Accessed October 1, 2018.

Bradshaw, S. (2001) Reconstructing Roles and Relations: Women's Participation in Reconstruction in Post-Mitch Nicaragua. *Gender and Development.* 9, pp. 79–87.

Carmack, R. M., Gasco, J. L. & Gossen, G. H. (2006) *The Legacy of Mesoamerica: History and Culture of a Native American Civilization.* New York, Routledge.

Castañaza, C. (2015) Hubo Advertencia Sobre la Tragedia. *Prensa Libre* website, October 4. Available at www.prensalibre.com/hubo-advertencia-sobre-la-tragedia. Accessed September 12, 2018.

Cernea, M. M. (1996) The Risks and Reconstruction Model for Resettling Displaced Populations. *World Development.* 25 (10), pp. 1569–1587.

Cupples, J. (2007) Gender and Hurricane Mitch: Reconstructing Subjectivities After Disaster. *Disasters.* 31, pp. 155–175.

D'Ans, A. M. (1998) *Honduras: Emergencia difícil de una nación, de un Estado.* Tegucigalpa, Renal Video Producción.

De las Casas, B. (2006) *Brevísima relación de la destrucción de las Indias.* Medellín, Editorial Universidad de Antioquia.

Donado, J. R. (2008) *Aporte de la Antropología Social a la investigación forense para casos de Desastres Naturales (Cantón Cuá, Municipio de Tacaná, departamento de San Marcos 2005–2006).* BA Thesis. Guatemala, Universidad de San Carlos de Guatemala.

Ensor, M. O. (2009a) Gender Matters in Post-Disaster Reconstruction. In: Ensor, M. O. (ed.). *The Legacy of Hurricane Mitch.* Tucson, University of Arizona Press, pp. 129–155.

Ensor, M. O. (ed.). (2009b). *The Legacy of Hurricane Mitch.* Tucson, University of Arizona Press.

Escobar, A. (1995) *Encountering Development: The Making and Unmaking of the Third World*. Princeton, Princeton University Press.

Ferguson, J. (1994) *The Anti-Politics Machine: "Development," Depoliticization, and Bureaucratic Power in Lesotho*. Minneapolis, University of Minnesota Press.

Fuentes, V. (2009) The Response of Civil Society to Hurricane Mitch. In: Ensor, M. O. (ed.). *The Legacy of Hurricane Mitch*. Tucson, University of Arizona Press, pp. 100–128.

Fuentes, V. y Guzmán, F. (1932) [1690] *Recordación Florida: Discurso Historial y Demostración Natural, Material, Militar y Política del Reyno de Guatemala*. Guatemala, Tipografía Nacional.

Gamarro, U. (2018) Cepal: Erupción del Volcán de Fuego desarticuló circuito económico. *Prensa Libre* website, July 12. Available at www.prensalibre.com/economia/evaluacion-de-la-cepal-por-impacto-de-erupcion-volcan-de-fuego-cuantificacion-de-dao-y-perdidas. Accessed September 10, 2018.

García-Acosta, V. (2018) Los Desastres en Perspectiva Histórica. *Arqueología Mexicana*. 149, pp. 32–35.

García-Acosta, V. (2002) Historical Disaster Research. In: Hoffman, S. M. & Oliver-Smith, A. (eds.). *Catastrophe & Culture: The Anthropology of Disaster*. Santa Fe, Nuevo México, School of American Research, pp. 49–66.

García-Acosta, V. (1993) Enfoques teóricos para el estudio histórico de los desastres naturales. In: Maskrey, A. (comp.). *Los desastres no son naturales*. Bogota, LA RED/Tercer Mundo Editores, pp. 128–133.

Garrido, K. C. (2007) *La Organización en Torno a la Prevención y Mitigación de Desastres en la Aldea El Volcán, Camotán, Chiquimula*. BA Thesis. Guatemala, Universidad de San Carlos de Guatemala.

Gómez, S. R. (2018) Hagamos la Diferencia: Se vale soñar. *Prensa Libre* website, October 9. Available at www.prensalibre.com/opinion/opinion/se-vale-soar. Accessed September 15, 2018.

Guatemala, Memory of Silence (1999) Guatemala, Comisión para el Esclarecimiento Histórico.

Guzmán Böckler, C. & Herbert, J. L. (1970) *Guatemala: Una interpretación histórico-social*. México, Siglo Veintiuno Editores.

Hewitt, K. (ed.). (1983) *Interpretations of Calamity: The Viewpoint of Human Ecology*. Boston, Allen & Unwin.

Holland, C. J. & van Arsdale, P. W. (1989) Aspectos antropológicos de los desastres. In: *Desastres: Consecuencias Psicosociales: La Experiencia Latinoamericana*, Section 14. México, Programa de Cooperación Internacional en Salud Mental. Available at http://helid.digicollection.org/en/d/Jph30/7.5.1.html#Jph30.7.5.1. Accessed October 1, 2018.

Holston, J. (1989) *The Modernist City: An Anthropological Critique of Brasilia*. Chicago, University of Chicago Press.

Jansen, K. (1998) *Political Ecology, Mountain Agriculture, and Knowledge in Honduras*. Amsterdam, Thela.

MacLeod, M. (1973) *Spanish Central America: A Socioeconomic History, 1520-1720*. Berkeley, University of California Press.

Marshall, J. S. (2007) The Geomorphology and Physiographic Provinces of Central America. In: Bundschuh, J. & Alvarado, G. E. (eds.). *Central America: Geology, Resources and Hazards*. London, Taylor & Francis.

Martínez Peláez, S. (2009) *La Patria del Criollo: An Interpretation of Colonial Guatemala*. Durham, Duke University Press.

Miner, Y. S. (2002) Determinación de vulnerabilidades temáticas en cuatro asentamientos humanos del área metropolitana de Guatemala, ante la amenaza de deslizamientos: Un aporte a la Antropología Urbana. BA Thesis. Guatemala, Universidad de San Carlos de Guatemala.

O'Keefe, P., Westgate, K. & Wisner, B. (1976) Taking the Naturalness Out of Natural Disasters. *Nature*. 260, pp. 566–567.

Oliver-Smith, A. (1999) What Is a Disaster? Anthropological Perspectives on a Persistent Question. In: Oliver-Smith, A. & Hoffman, S. M. (eds.). *The Angry Earth: Disaster in Anthropological Perspective*. New York, Routledge, pp. 18–34.

Oliver-Smith, A. & Hoffman, S. M. (eds.). (1999) *The Angry Earth: Disaster in Anthropological Perspective*. New York, Routledge.

Orozco, A. & Ramos, J. (2016) Edil de Santa Catarina Pinula es Ligado a Proceso. *Prensa Libre* website, August 30. Available at www.prensalibre.com/guatemala/justicia/edil-de-santa-catarina-pinula-es-ligado-a-proceso. Accessed September 15, 2018.

PAHO (Pan American Health Organization). (1998) Impact of Hurricane Mitch on Central America. *Epidemiology Bulletin*. 19 (4), pp. 1–13.

Paolisso, M., Gammage, S. & Casey, L. (1999) Gender and Household-Level Responses to Soil Degradation in Honduras. *Human Organization*. 58 (3), pp. 261–273.

Paredes, E. & Morales, S. (2018) Víctimas de la erupción del Volcán de Fuego no ven apoyo del Gobierno. *Prensa Libre* website, September 3. Available at www.prensalibre.com/ciudades/escuintla/erupcion-volcan-de-fuego-guatemala-san-miguel-los-lotes-3-de-junio-de-2018. Accessed September 4, 2018.

Patzán, J. M. (2017) Así Luce El Cambray, La Colonia Fantasma después de dos años de la tragedia. *Prensa Libre* website, October 1. Available at www.prensalibre.com/ciudades/guatemala/cambray-2-tragedia-que-sobrevivientes-despues-tragedia-colonia-fantasma. Accessed September 24, 2018.

Pérez de Lara, O. (1987) El Desarrollo de la Antropología en Guatemala: Necesidades y Perspectivas. *Boletín de Antropología Americana*. 16, pp. 113–134.

Ramírez Cruz, A. L. & Rodríguez Herrera, A. (1993) Algunas reflexiones sobre el desarrollo de la Antropología en El Salvador. *Cuadernos de Antropología*. 9, pp. 37–47.

Ramos, J. (2016) Capturan a Antonio Coro por la tragedia de El Cambray 2. *Prensa Libre* website, January 15. Available at www.prensalibre.com/guatemala/justicia/detienen-a-antonio-coro-excalde-de-santa-catarina-pinula. Accessed September 12 2018.

Snarr, D. & Brown, E. (1980) User Satisfaction with Permanent Post-Disaster Housing: Two Years After Hurricane Fifi in Honduras. *Disasters*. 4 (1), pp. 83–91.

Stonich, S. (1993) *"I Am Destroying the Land!" The Political Ecology of Poverty and Environmental Destruction in Honduras*. Boulder, Westview Press.

Tax, S. (1953) *Penny Capitalism: A Guatemalan Indian Economy*. Chicago, The University of Chicago Press.

Wagner, R., Hempstead, W., Villegas, B. & von Rothkirch, C. (2001) *The History of Coffee in Guatemala*. Bogotá, Villegas Editores.

Way, J. T. (2012) *The Mayan in the Mall: Globalization, Development, and the Making of Modern Guatemala*. Durham, Duke University Press.

4 Thinking through disaster

Ethnographers and disastrous landscapes in Colombia

Alejandro Camargo

Conversations

This chapter endeavors to present the current state of anthropological research on disasters in Colombia. It does so through an analysis of the experiences of those conducting ethnographic research in different places across the country, which have been affected by disaster. Unlike other Latin American countries, such as Mexico, Colombia does not have a distinct field of academic inquiry known as the Anthropology of Disaster. Colombian anthropology departments do not offer courses in that field, anthropologists have not framed their work within that intellectual tradition, publications on this matter are scant, and the history of the discipline does not present evidence of the formation of an academic community recognizable under that label. But the fact that such a tradition is underdeveloped does not necessarily mean that anthropologists have never conducted fieldwork in areas affected by disasters, or that Colombia is a not a disaster-prone country. Quite the contrary, as this chapter will show, this country has suffered a number of disasters affecting different places and people. Furthermore, those disasters have been part of the ethnographic experience of various anthropologists. This chapter examines those experiences and reflects on how disasters have shaped anthropological analysis in Colombia.

Since tracing a genealogy of the transformation and trends of anthropological approaches to disaster is not possible for the reasons explained already, I have chosen a different methodological path. This chapter draws on a number of conversations I have had with anthropologists who either have studied some aspect of a disaster, or have unexpectedly experienced the aftermath of a disaster during their ethnographic research on other subjects. In both cases, the disaster was not only an event which remains stored in people's memories, archives, or other documents. The disaster is, primarily, a material phenomenon with an obvious and enduring imprint on the landscape. Therefore, the different ethnographic experiences presented here have this in common: that the materiality of a disastrous landscape significantly affects the ways in which anthropologists understand and approach their research. This chapter devotes particular attention to how the materiality of disaster, and the configuration of the things and elements that constitute those landscapes, triggers ethnographic curiosity. Virginia García-Acosta has long argued that disasters, as processes, "can serve as the thread on which we, through ethnography, can weave many cultural histories" (García-Acosta, 2002: p. 65). The same can be said about disastrous landscapes insofar as they embody socioenvironmental histories, and, at the same time, provide an opportunity for anthropologists to weave histories with them. The disaster and its materiality are the medium through which the ethnographers that participate in this chapter have considered broader anthropological questions.

Thus, while in most anthropological works the voice of colleagues is second-ary, and usually cited as "personal communication", in this chapter those conver-sations are center stage, as those ethnographers are the main interlocutors in this analysis. Existing anthropological works on disasters in Colombia are mostly lim-ited to a small body of literature and a larger set of unpublished theses. This chap-ter considers both of those sources, and also some additional nonanthropological documents that help to contextualize the main discussions. The rest of the chap-ter is organized into seven parts. The first and second parts provide an overview of the major disasters that have occurred in Colombia, as well as the main concerns of Colombian anthropology, respectively. The next section develops the concept of a disastrous landscape. The fourth section explains how the Armero tragedy, a disaster produced by the eruption of a volcano, became a scenario for the con-stitution of the only research group of ethnographers ever created in Colombia to study a disaster. Then I examine how the instability of disaster-affected land

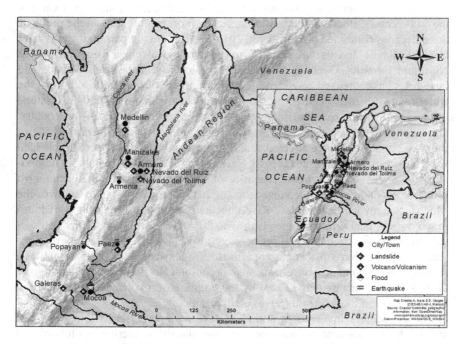

Map 4.1 Map Colombia. Case studies and main areas mentioned

shaped the work of ethnographers interested in issues such as property rights and territorial zoning. Then I will analyze how an avalanche intersected with the long tradition of indigenous studies in Colombian anthropology. In the last section, I suggest some potential ways to develop an Anthropology of Disasters in Colombia which is attentive to the materiality of the landscape and in conversation with other debates and fields such as history, sociology, and geography.

A country of disasters

Disasters have been a significant part of the socioenvironmental history and the territorial configuration of Colombia. In fact, the Global Facility for Disaster Reduction and Recovery (GFDRR) describes Colombia as having the highest recurrence of extreme hazard events of any country in South America (GFDRR, 2018). In addition, between 1970 and 2011, more than 28,000 events in Colombia have been reported (World Bank, 2016). This figure should not suggest that disasters were uncommon prior to 1970. Historians, geographers, and archaeologists have called attention to the occurrence of disasters in the past, and the significance of these events to the understanding of specific historical moments (Espinosa Baquero, 1997; Sarabia Gómez et al., 2010; Serna Quintana, 2011; Therrien, 1995).

Hydrometeorological disasters have been the most common, floods and landslides being the cause of the greatest percentage of disaster-related loss of life and housing destruction (Campos García et al., 2011). In fact, in the first decade of the 21st century, floods and landslides affected more people than in the previous three decades combined (World Bank, 2013). This is due, in part, to the country's tropical climate and its vulnerability to meteorological phenomena such as El Niño and La Niña. Over the last decade, for instance, catastrophic floods and landslides have wreaked considerable havoc in many places. In this context, the water-related disasters caused by the La Niña phenomenon in 2010–2011 stood out as one of the most devastating events in the recent history of the country. During this extreme weather event, about four million people (around 8.5% of the total population) were affected, and nearly one million hectares of arable land were destroyed (Sánchez, 2011: p. 8). Between 2017 and 2018, a number of landslides also created catastrophic scenarios of suffering and pain. One of the most dramatic events was the Mocoa landslide of 2017. Intense rainfall overflowed the Mocoa, Sangoyaco, and Mulata Rivers located within the Department of Putumayo, killing nearly 320 people (Ishizawa, 2018; Telesur, 2017). Other geological events of less frequency – yet of great impact – such as volcano eruptions and earthquakes have also configured disastrous landscapes and devastating loss. Among the most memorable events are, chronologically, the Popayán earthquake (1983), the Armero tragedy caused by the eruption of a volcano (1985), and the Paez (1994) and the Armenia (1999) earthquakes.

These disasters involved profound socioenvironmental transformations and shaped the evolution of governmental disaster and risk reduction institutions and strategies. In 1984, the government created the Fondo Nacional de Calamidades

(National Fund for Calamities). In 1988, after the Armero tragedy, the Sistema Nacional para la Prevención y Atención de Desastres (National System for Disaster Prevention and Attention) came into being as a synergic combination of state agencies. These agencies worked together to create policies and to address future disasters (UNGRD, 2018). In the aftermath of the La Niña floods in 2010–2011, public debate over climate change and climate adaptation shaped the ways in which disaster was understood. As a consequence, the government created the Fondo de Adaptación (Adaptation Fund) in December 2010, which, under the motto of climate adaptation, was in charge of funding the economic and infrastructural recovery of the areas affected by the La Niña floods. The idea of adaptation, as applied to disaster government, meant the incorporation of a longer-term perspective beyond prevention and response. In addition, the government created in 2011 the Unidad Nacional para la Gestión del Riesgo de Desastre (National Unit for Disaster Risk Management) to reinforce disaster risk management policies nationwide.

Many pages could be devoted to explaining how the aforementioned disasters were far from "natural". Increasingly, multiple political, economic, and social processes that underlie and precede the production of disaster become evident. Yet the long tradition of critical disaster studies in anthropology and other disciplines has already provided rich analytical tools to unveil the power relations that are embedded in floods, earthquakes, and other calamities (i.e., Oliver-Smith, 2002; O'Keefe et al., 1976; Faas, 2016). These tools also embody the idea of risk as socially and collectively constructed (Douglas & Wildavsky, 1982; García-Acosta, 2005). My concern here is not about the politization of disaster, but about how these abrupt socioenvironmental phenomena have shaped the intellectual interest of anthropologists.

Invisible disasters

Although Colombia is a disaster-prone country with a significant history of disasters that has dramatically changed the lives of thousands of people, anthropologists have devoted scant attention to these rather evident and transformative events. Consequently, an internationally recognized Anthropology of Disasters and Hazards has not found a niche in Colombian academia thus far. This is evident, for instance, in the two-volume book *Antropología hecha en Colombia* (Anthropology made in Colombia), which was published in 2017 by the Colombian Institute of Anthropology and History. This collection compiles a series of articles that were published at different times and outlets, but that together provide a picture of the main concerns of Colombian anthropology. A very prominent aspect of this overview is anthropologist's longstanding interest in indigenous people, movements, and indigenism. These topics not only constitute specific chapters in the book, but they also pervade other sections on ethnology, urban anthropology, ethnicity and multiculturalism, and historical anthropology. This interest in indigenous people has been present since the origins of anthropology as a discipline in Colombia, when the first anthropology school was created under the name of Instituto Etnológico Nacional (National Ethnological Institute) in 1941. As Roberto Pineda points out, the Institute had a particular teaching focus on the study of ethnology, which at the time was tantamount to Americanism (Pineda, 2004: p. 61).

Other subfields of traditional interest depicted in *Antropología hecha en Colombia* include Afro-Colombian communities, social movements, violence and armed conflict, and gender and sexuality. None of these themes include the study of disasters and hazards, nor do these words appear in the index section of the two volumes. The same reticence can be found in the edited volume *Antropologías en Colombia: tendencias y debates* (Anthrolopogies in Colombia: trends and debates) by Jairo Tocancipá (2016). At the regional level the situation is not different. In the review of the history of anthropology in the Cauca region compiled by Tabares and Meneses (2016), the study of disasters is never mentioned. This is particularly ironic since the Cauca region underwent the tragic consequences of the Páez earthquake and, as will be demonstrated in this chapter, some anthropologists conducted research in that context.

Colombian anthropology has gone through a process of diversification and institutional growth. In 1952, the Instituto Etnológico Nacional (National Ethnological Institute) merged with the Servicio Arqueológico Nacional (National Archaeological Service) to give rise to the Instituto Colombiano de Antropología (Colombian Institute of Anthropology), known today as the Colombian Institute of Anthropology and History -ICANH. It was not until the 1960s and early 1970s, however, when the first four undergraduate programs in anthropology appeared. These programs were expanded in the late 1990s to include eight undergraduate, five masters, and three doctorate programs. This institutional growth occurred together with a transformation of the dominant themes and research interests. It is not my intention to review this intellectual evolution here since the cited edited volumes, among others, have done this comprehensively. Yet, it is must be noted that the predominant interest in ethnic groups had a particular resonance in the 1970s. At this time, anthropologists were theoretically engaged with Marxism since this tradition proved a useful framework to understand exploitation, acculturation, and the marginality of indigenous groups, rural societies, and the urban poor (Bernal Gamboa, 2016). This Marxist approach lost its predominance in the 1990s, when an "anthropology of modernity" emerged in Colombia in conversation with international debates in cultural studies, postcolonial theory, and subaltern studies (Restrepo, 2016). Colombian anthropology has diversified institutionally, thematically, and theoretically as disasters have continued to occur and to wreak their havoc throughout the country. Paradoxically, these two trends have barely intersected. As a consequence, disasters have been invisible in Colombia anthropology. Yet disasters have provided the material substratum, the milieu through which researchers reflect on other topics beyond the disaster. When anthropologists meet disasters in the field, they do not merely meet with a catastrophic event. They also meet with a disastrous landscape whose materiality significantly shapes their ethnographic experience. Before delving into these experiences, I will briefly explain the concept of disastrous landscapes.

Disastrous landscapes

Earthquakes, floods, and mudslides are not merely hazardous events that result in disasters and social calamities. These phenomena are also material disruptions which involve the restructuration of the surface of the earth and the redistribution of elements and things upon it. These physical transformations are abrupt and inevitably surpass our abilities to control our material environment. Disasters are dramatic in their impact on people's lives and their perceptions of the world. Diego Cagüeñas argues that disasters expose our limitations because we are unable to undo what happened. The inevitable, the finite, and the irreversible shape the experience of disaster, but at the same time they pave the way to the reconstruction of life (Cagüeñas, 2013: p. 224) in the aftermath of the crisis. Therefore, post-disaster reconstruction and climate adaptation are nothing more than a quest for restoring life in the midst of ruination and death. Yet these tensions between destruction and reconstruction, and between death and life, are as much a matter of human experience and meaning as they are material causes of landscape transformation.

Girot and Imhof define landscape as a "cultural artifact," "a construct resulted from the belabored shaping of terrain and the making of place; it has, in fact, very little to do with the ideal of an untouched wilderness" (Girot & Imhof, 2016: p. 7). Hirsch (1995) cautions that some conceptualizations of landscapes as cultural images or pictorial ways of representing surroundings (such as those of geographer Denis Cosgrove) may portray landscape as a static reality. Rather, Hirsch proposes an understanding of landscape as a cultural process and more closely related to what happens in the everyday life. In this way, landscapes are shaped by ideas, meanings, and practices. As such, they are historically dynamic places where the record of the lives and work of past generations endure (Ingold, 1993). Thus, landscapes reflect, and are a part of, change. Yet they also incorporate continuity and connect the past, the present, and the future (Stewart & Strathern, 2003: p. 4). As Ingold points out, "the landscape is always in the nature of a 'work in progress'" (1993: p. 162).

But what does disaster add to these conceptualizations of landscape? Disasters are spatial, temporal, and material phenomena (Baez Ullberg, 2017: p. 2), and occur along a historical pattern of vulnerability (García-Acosta, 2004: p. 130). Therefore, disastrous landscapes reflect meanings, practices, inequalities, and ideas. But they also materialize the disruptive power of geological and meteorological phenomena. Certainly, in times of climate change and environmental degradation on a global scale, these phenomena are increasingly seen as anthropogenic. Nevertheless, disasters such as earthquakes and floods also remind us of, as Cagüeñas (2013) has observed, our own limitations as humans. It can be said that disastrous landscapes are not only about human perception and vulnerability, but also about the materiality of biophysical processes and the role of that materiality in the production of meaning.

Stewart and Strathern (2003) have pointed out that ethnographers, through their field experiences, realize how perceptions, values, and memories attached to landscapes make them sites of historical identity. One could argue, therefore, that an anthropological analysis of landscape emphasizes how landscapes are socially produced and that ethnography helps us understand that production process through description and interpretation (Hirsch, 1995). While most anthropological work would focus on the people who inhabit and work those landscapes, in this chapter I will concentrate on how ethnographers experience those disastrous landscapes and the ways in which those material formations shape their anthropological work. Disastrous landscapes are spaces of "ethnographic revelation", where "unanticipated, previously unconceivable things become apparent" (Henare et al., 2007: p. 2). Ingold maintains that for archaeologists and native dwellers the landscape narrates its own story (1993: p. 152). They *read* the landscape in order to understand the lives and times of those who inhabited and transformed the material world. Ethnographers experience the landscape in a similar fashion. They read the landscape not only to understand the past, but also to project their intellectual work onto the future. Although unanticipated and unconceivable things have long been part of the ethnographic experience, the destruction and devastation materialized in the landscape makes that experience a more shocking, transformative, and striking one.

The idea of disastrous landscapes as spaces of ethnographic revelation, therefore, stands in sharp contrast with the very definition of disasters as revelatory events. Disasters "accentuate" the intricate relationships among the environment, politics, culture, and power (Claus et al., 2015: p. 291) in their destructiveness. Barrios (2017) reminds us that the revelatory nature of disasters is contingent on the positionality of the beholder who then problematizes the disastrous phenomena. In the cases discussed in the following sections, the beholders are the ethnographers and the disaster is not an object of reflection per se. Rather, it is a material means *through* which ethnographers enter other worlds. The revelatory character of disasters is at the same time a source of ethnographic revelation. Thus, disastrous landscapes are *things*, in the sense suggested by Henare et al. (2007), that instigate meaning during the ethnographic encounter. The positionality of the ethnographer, as a beholder, has a material dimension because he/she "needs to be there, to experience the landscape through the sensual and sensing body, through his or her corporeal body" (Tilley & Cameron-Daum, 2017: p. 4). In the next sections, I will explain the meanings and experiences that various ethnographers working in different regions of Colombia have built in their encounters with disastrous landscapes.

Armero: a theory of the world

On 13 November 1985, an eruption of the Nevado del Ruíz volcano triggered one of the most dramatic and notable tragedies in Colombia. Volcanic materials melted glaciers and coursed into nearby rivers thereby affecting several settlements in its pathway. The town of Armero, in the Department of Tolima, was one of the most severely affected communities. Official figures reported that around 25,000 people were killed, many others were never found, and thousands lost everything. Images of this tragedy circulated worldwide, along with criticism of the government's role in the prevention of this tragedy and its failure to assist the victims adequately.

Almost 20 years after this tragedy, in 2007, Luis Alberto Suárez Guava, a professor of ethnography in the Department of Anthropology at the Universidad Nacional de Colombia, brought a group of undergraduate students on a fieldtrip to Armero. According to Luis Alberto Suárez Guava, the goal of this trip was to expose the students to a particular context without a previous research agenda or theoretical framework. They visited Armero on the commemoration day in order to try to comprehend what was going on there, and then to turn that experience into an ethnographic writing exercise. This was the beginning of what later came to be known as the "Etnografía y memoria de Armero" group (Ethnography and Memory of Armero), the only research group ever created in Colombia to study a disaster from an anthropological perspective. Mónica Cuéllar was one of those students who went to Armero in the context of the ethnography course. For Cuéllar, the first encounter with this place was a very striking experience because the ruins, the memorial sites, and the landscape in general are still fraught with an atmosphere of death, although they also come to life in unexpected ways. Every year on the commemoration day former inhabitants and relatives of those who perished in the tragedy gather together in the town. The attendees have lunch at the places where their homes were built. Then they celebrate a mass where the town's church used to be and also partici-pate in other practices of collective memory, such as a "rain of flowers" that is made to fall over the ruins as a way to represent the avalanche. The combination of a devastated landscape and commemorative cultural practices led Cuéllar, together with other four students, to write their undergraduate theses on this particular place.

For more than a year, Cuéllar visited Armero on several fieldtrips. Her intention was to understand the entanglement between stories about the disaster and stories of love and heartbreak (Cuéllar, 2010). For Cuéllar, her project and the whole work of the Armero group was an ethnographic journey in search of a "theory of the world" that would emerge out of the disaster. By "theory of the world," she refers to the ways in which the people who experienced the disaster crafted conceptual, material, and affective constellations out of the pieces of their shat-tered world in order to make sense of it, and reinhabit it, in the aftermath of the avalanche. This theory of the world is grounded in a landscape which comprises geographical formations such as volcanoes and mountains, as well as buried objects such as *guacas* (ancient indigenous graves), gold, and emeralds (Suárez Guava, 2009). The group found that for many of the victims, the disaster meant "the end of the world" (Uribe de Zuluaga, 2009). As such, the disaster constituted a thorough and dramatic transformation of their lives. The Armero group sought to understand those theories of the world and the meanings, memories, geological formations, and things that sustained such cosmological order. This experience rendered ethnography, according Luis Alberto Suárez Guava, an epistemological option rooted in anthropology, rather than just a method.

This approach pervaded the way in which the group understood ethnography and anthropological analysis. The experience in Armero was, for Cuéllar and the other students, a remarkable and transformative moment in their training as anthropologists and human beings. "I became an anthropologist in Armero", said Cuéllar during our conversation. By this, she meant that her vocation and her anthropological perspective took shape in the field as she and the others tried to understand the intricacies of people's worldviews at that conjuncture. Their approach was not to make the field a terrain for the validation of theories, but an arena for theoretical and empirical exploration. In that sense, their priority was not to understand what the Armero case tells in terms of a specific anthropological debate, but how the victims of this catastrophe explain their world in the aftermath of such a disaster.

The intention of the group was not to conduct ethnographic research in Armero in order to produce specific academic outcomes, but to understand what ethnography is. As Andrés Ospina, another member of the group, explained to me: "we did not have specific theoretical guidelines or frameworks to approach the field; we just went there to see what was going on and that decision had a tremendous value from an ethnographic viewpoint". Thus, the work of this group has been mainly preserved as academic theses and only a few published articles. This work explores topics such as witchery, commemoration, memory, child rearing, and emotions through the disaster (e.g., Acero Pulgarín, 2010; Buitrago Ospina, 2012; Cuéllar, 2011; Ospina, 2013; Suárez Guava, 2008). Eventually, the group considered the possibility of putting together a book, but the idea did not find sufficient support. As Cuéllar pointed out, the Armero group was marginal in the Anthropology Department, as the bulk of the mainstream research done at that time was mostly oriented to issues such as social mobilization, armed conflict, and violence. Talking about a disaster was not seen as relevant at that particular moment.

Fluid ethnographic grounds

Ethnographic work in disastrous landscapes has also considered the instability and fluidity of certain terrains, surfaces, soils, lands, and water which have also opened new avenues for anthropological research in Colombia. Anthropologist Lorenzo Granada currently studies the "meanings of mud" in Armero. Mud, he explained to me, "is death, but it's also life as it is very fertile for agriculture and vegetation in general." During the disaster, mud erased human identities by covering bodies and homogenizing the landscape. Years later, this very same mud created the ecological conditions for the expansion of agricultural enterprises, which found those volcanic soils very fertile and an ideal terrain for cultivation. Mud, therefore, has been a central element in both the recollection of the disaster and the reconstruction of Armero. Nevertheless, the concept of disaster, for Granada, is insufficient to explain the life of mud in the aftermath of the catastrophe. Conversely, this concept is useful to analyze how this calamity created an empty space by erasing a town from the landscape. This spatial erasure posed serious challenges to the ways in which the affected area was seen and represented geographically. For Granada (2016), an empty map and a fragmented landscape intermingled with stories of indigenous' skulls, spells, and rubble. These entanglements of material and immaterial things highlight the idea of volcanoes as an "overturn of earth" associated with stories of guacas. This linkage between disastrous geological formations and ancient human records is very common among the people of the Armero area. It is a fundamental part of their cosmology (Suárez Guava, 2009: p. 379).

While in Armero mud has settled, in other places mud and the terrain are unstable and always in motion. That movement creates constant difficulties and makes social relations unstable as well. Marie McDonald conducted research in Manizales, a city which has repeatedly suffered the effects of mudslides. McDonald decided to conduct her dissertation research in this city because the materiality and dynamics of mudslides did not follow the conventional logic of a disaster: an event whose beginning and end can be traced over time. Mudslides in Manizales occur constantly and although they cause trouble and present risk, they have not reached a catastrophic level. Consequently, mudslides become events that are an intrinsic part of the spatial and political configuration of the city. Since they do not lead to spectacular disasters, they gain little attention in the media. The materiality of muds along with the topographic configuration of Manizales also create a space of governmental intervention. It is at this intersection of mudslides, topography, and politics where McDonald configures her research goals.

The Mocoa landslide in 2017 also posed unexpected issues to ethnographers working in that area. Mocoa is the capital city of the Department of Putumayo where a number of anthropologists have long conducted research on various topics such as multiculturalism and the state (Chaves, 2010), peasant movements (Ramírez, 2001), and armed conflict and resistance (Cancimance, 2015), among others. Kristina Lyons has also conducted ethnographic research in Putumayo since 2004. Her work was initially focused on the environmental and social impacts of glyphosate, as this area has been blighted by coca-leaf cultivation for drug markets, and its attendant armed conflict and entanglements with militarized antidrug policy. Later on, her work with the rural communities of Putumayo led her to examine the cultivation of life in the midst of death and poison. Specifically, she has studied soils as key actors in nature-society relationships. The landslide of 2017, however, presented other possibilities to understand the materiality of those relationships. This disaster involved a particular connection between rivers, rocks, topography, and semi-urban and rural spaces. While considering the micropolitics of soils, the disaster led Kristina Lyons to observe other elements such as watersheds on broader scales.

Lyons' recent involvement with the disastrous landscape of Mocoa calls attention to a methodological challenge of relevance to the anthropological study of disasters. If disasters are the product of the interactions between "physical, biological, and sociocultural systems" (Oliver-Smith & Hoffman, 2002: p. 4), then ethnography is insufficient to account for the complexity of a catastrophe. For Lyons, "rivers have memory" (2018), which means that disasters such as landslides may occur when rivers return to their previous courses. Accordingly, it is important to understand how rivers function and what geomorphological and hydrological processes shape the ways in which water, rocks, and land flow. Specifically, Lyons points out, geomorphology help us "read" the landscape, the rocks, and the history of landslides in the area.

My own ethnographic work on disastrous floods in northern Colombia has demanded a diversification of methods well beyond the social sciences. For instance, in order to comprehend the catastrophic effects of the La Niña floods in 2010 and 2011, we need to understand the hydrological processes that underlie the production of floods. By examining those hydrological processes, we realize that floods are not exceptional events. Rather, floods can also be deemed as constitutive phenomena of the geographical configuration of a landscape. When floods become catastrophic, it is because some socioecological transformations have taken place which have enhanced vulnerability. Wetland drainage, for instance, increases the likelihood of catastrophic floods when rainfall intensifies. In order to understand these landscape transformations, however, it is necessary to put together oral histories, archival records, aerial photographs, and historical maps, among other resources. Therefore, my experience with floods and their material effects on rural landscapes demanded a broader methodological approach which had ethnography as a key component, but which also relied upon the assistance of hydrology, archival work, climatology, and cartography (Camargo, 2016; see also Camargo & Cortesi, 2019). This commixture of methods illuminates a different interpretation of floods. While the Colombian government explained the La Niña floods of 2010–2011 as a consequence of climate change, I found that these floods were yet another chapter of a longer history of landscape transformation, agrarian development, and infrastructural failure (Camargo, 2016).

Ethnographic encounters with disastrous landscapes and "shaky grounds" (Zeiderman, 2012) foster methodological diversification, even within anthropology. This is what happened to Meghan Morris while she was conducting research in Medellín. She was interested in property relations in a mountainside neighborhood. The inhabitants of this neighborhood were forcefully displaced from the countryside and, consequently, obliged to settle in the margins of the city. While walking with people around the neighborhood and talking about the experience of living in that place, she began to hear stories about a disaster. Torrential rains had triggered a landslide killing many people and destroying homes. Morris explains that for the people of this neighborhood this disaster also meant a change in their relationship with property and their everyday spaces. The landslide paved the way for the implementation of a new regime of risk governance in the area which was based upon the denomination of places as high-risk zones. This in turn has created uncertainty among people regarding the future of property relations in the neighborhood. For Morris, this linkage between property and disaster was unexpected: "I found this disaster without looking for it. . . . It was a classical ethnographic encounter." This encounter had significant methodological implications for her work. On the one hand, she employed the approach of *walking along with people* as a research method. Although some anthropologists have elaborated on the ethnographic importance of walking (Ingold & Vergunst, 2008), the "go-along" research tool (Kusenbach, 2003), and "moving along with people" (Horisberger, 2018), Morris' experience shows the relevance of that method for the anthropological study of disastrous landscapes. On the other hand, this encounter made her more attentive to the materiality of the landscape and opened up new avenues of inquiry regarding the relationships between property, soils, and the physical qualities of the ground.

Ethnicity and disaster

In 1994, another landslide shook the ordinary lives of hundreds of people. In early June, an earthquake unleashed an avalanche of mud and rocks and severely affected a number of villages located in an indigenous territory along the Páez River in Southwestern Colombia. Nearly 1,300 people died, others disappeared, and many houses were destroyed. This disaster mobilized all sorts of humanitarian aid, state agencies, and other forms of solidarity, as would be expected in a situation like this. But this calamity also shaped the research agenda of some anthropologists at the Universidad del Cauca, which is located in the same region. The Anthropology Department at the Universidad del Cauca was established in 1970 and was one of the four pioneer programs together with Universidad de los Andes (1964), Universidad Nacional de Colombia (1966), and Universidad de Antioquia (1967). The 1970s, as mentioned earlier, were a crucial period for the development of anthropology as a discipline in Colombia. At that time, a militant anthropology in support of the nascent indigenous movements took shape.

A part of this community of anthropologists joined the Movimiento de Solidaridad con los Pueblos Indígenas (Solidarity with Indigenous People Movement). This movement of intellectuals worked closely with the Consejo Regional Indígena del Cauca (Regional Indigenous Council of Cauca). Through their joint effort with indigenous movements, these scholars sought to change Colombian anthropology. According to them, the discipline was in danger of reproducing the colonial approach and the unequal relations with interlocutors that characterized European anthropology. As a consequence, researchers become the sole holders of knowledge, and their communities of origin the only beneficiaries of it (Caviedes, 2007). The Movimiento de Solidaridad con los Pueblos Indígenas, according to Caviedes, claimed a collaborative anthropology having solidarity as one of its main principles. Consequently, these anthropologists played central roles in the formation of indigenous movements, the implementation of intercultural and bilingual education projects in indigenous communities, and the development of politically committed research methods (Caviedes, 2002, 2007). This militant anthropology further sought to challenge cultural hegemony, the historical inequality and marginality of indigenous and campesino people, and the dominant narrative of national identity, which deemed indigenous people as backward (Jimeno, 2018). By doing so, anthropologists aimed at building a new state and reconfiguring a more inclusive nation (Bernal Gamboa, 2016).

These intellectual and social antecedents of anthropology at the University of Cauca had continuity in the disaster of 1994. Hugo Portela, a Professor at this university, recounted that a number of scholars from different programs at his institution mobilized in the aftermath of the catastrophe. These scholars were particularly engaged in the process of relocation of the affected people. Portela described a very memorable experience when he and anthropologist Herinaldy Gómez accompanied indigenous people from the Páez community (also known as the Nasa people) to walk together across the devastated landscape. They visited the most meaningful and sacred places, undertook healing rituals, and performed practices of collective memory. According to Portela, their role as anthropologists was not so much to produce academic knowledge as to collaborate and contribute to the recovery of indigenous communities. Walking together across the landscape was, as in Meghan Morris' ethnographic experience, a method to respond to the disaster and "heal" the wounds materialized in a landscape of destruction. The disastrous landscape of the Páez river basin, therefore, constituted a venue for the revitalization of a collaborative anthropology.

Shortly thereafter, the collaboration among indigenous peoples, anthropologists, and other scholars converged in the creation of a state-led organization called Nasa Kiwe. The Colombian government initiated this organization to direct the reconstruction of the villages and communities that were affected by the landslide. To that end, the government appointed Gustavo Wilches-Chaux, who is one of the most renown disaster specialist in Colombia and Latin America, as the director of Nasa Kiwe. This organization also functioned as a response to previous disaster management approaches to take into consideration the cultural and social particularities of the various ethnic groups. Thus, the Colombian government created Nasa Kiwe as a way to acknowledge that the catastrophe mostly affected indigenous people, and that this aspect of the disaster demanded a different approach. To this end, anthropologist Herinaldy Gómez and sociologist Carlos Ariel Ruiz conducted a participatory research project to examine the spatial and territorial practices, meanings, and memories of indigenous peoples which were incorporated in the disastrous landscape. This research was later published in 1997 as what seems to be the only anthropological monograph on disasters written by Colombian scholars: *Los Paeces: gente territorio, metáfora que perdura* (*The Paeces: territory people, metaphor that endures*). For the authors, the disaster constituted yet another chapter in a longer history of cultural and ethnic survival. This book has a special emphasis on how the Paeces explain and understand hazards and disasters within their own cosmology, and it analyzes the places of these worldviews in the broader construction and meanings of their territory and landscapes. Therefore, this study can be considered as an anthropological work about geosocial relations and cosmologies in the aftermath of disaster. Wilches-Chaux (1997) observed that this book was published at a time when scholars and practitioners predominantly approached disasters from the perspective of emergency and management. Henceforth, disaster would be seen as multidimensional, historical, and social phenomena. For Wilches-Chaux, *Los Paeces: gente territorio, metáfora que perdura* is an example of how the social sciences might engage with risk management programs.

Other anthropologists, who were previously involved in ethnographic research with the indigenous peoples of the Cauca area, also contributed to the understanding of the aftermath of the disaster. In 1995, LA RED (Social Research Network for Disaster Prevention in Latin America) published a special issue on the Páez River disaster in its journal *Desastres y Sociedad* (*Disasters and Society*). Anthropologists María Teresa Findji and Víctor Daniel Bonilla, who were also part of the development of the militant anthropology in the late 1970s and 1980s, collaborated in that volume. In their article, Findji and Bonilla (1995) explained how a combination of a history of mobilization, people's spatial dispersion, and territorial conflict since colonial times shaped the experience of the disaster in 1994.

The relocation process after the landslide particularly caught the attention of some anthropologists. North American anthropologist David Gow (2008) examined how the aftermath of the disaster and the relocation project created a scenario where different notions of development and social mobilization intersected. He also coauthored an article with Joanne Rappaport (Rappaport & Gow, 1997), whose ethnographic work with the Nasa people, both before and after the disaster, has been very influential in Colombian anthropology. To date, a number of anthropologists have written academic theses which focus on this disaster (e.g., Meneses Lucumí, 2000; Quiceno Montoya, 2018; Ramírez Elizalde, 2013). Through the landslide of the Páez River, these works examine topics such as cultural change, space-society relationships, and development. The disastrous landscape of the Páez River was instrumental for the elaboration of that analysis, but it was not the source of theoretical reflection. In fact, the book *Historia de la Antropología en el Cauca* (Tabares & Meneses, 2016) does not mention the study of disasters in its review. Rather, the disaster constituted a venue to further develop the longstanding interest in ethnic studies within anthropology.

Bridges and paths

This chapter presented a reflection on the ethnographic experiences of anthropologists who have encountered disastrous landscapes in their research projects in Colombia. These ethnographic encounters with an unexpected context of material devastation and social suffering inevitably shape academic agendas in different ways. However, these experiences are illustrative of the fact that encounters with disastrous landscapes have not always led to an engagement with the Anthropology of Disasters and Risk. Disasters have been a lens through which ethnographers have explored other themes of anthropological interest such as development, ethnicity, property relations, and human-nature relationships. Although these thematic connections are significant, the engagement with disastrous landscapes has had methodological and subjective ramifications as well. Landscapes of devastation and destruction have shaped anthropologist's reflexivity, ethnographic training, political commitment, and sensorial approaches to the field.

The analysis of these experiences has three implications for an Anthropology of Disaster in Colombia. First, disasters not only affect those who inhabit disastrous landscapes. Disasters also affect those who conduct research and who happen to have a connection with those places, people, and landscapes. Oliver-Smith (1999: p. 20) defined disaster as "totalizing events" since "all dimensions of a social structural formation and the totality of its relations with its environment may become involved, affected, and focused" as the disaster unfolds. The scope of these totalizing events would be broadened by an analysis of how ethnographers *read*, in Ingold's terms (1993), disastrous landscapes and how they relate their own personal stories. A substantial part of the literature on the Anthropology of Disasters focuses on the different "dimensions of a social structural formation", but little has been said about the meanings, stories, and moments of "ethnographic revelation" that emerge from the personal experience of anthropologists.

The second implication concerns the importance of the materiality of disastrous landscapes for the anthropological study of disasters. Doug Henry (2005) argues that anthropology, through its holistic approach, offers disaster studies comparative contextual, and cross-cultural perspectives. These perspectives situate disasters within broader political, economic, social, and environmental relationships and emphasize issues of vulnerability and culturally informed adaptation and responses. A large body of research in anthropology has been undertaken along those lines. However, the Anthropology of Disaster can greatly benefit from current discussions in other anthropological subfields and in other disciplines, which engage more closely with the materiality of the landscape and the environment. For instance, research on ruins and ruination (e.g., Gordillo, 2014; Stoler, 2013; Tsing, 2014), may offer rich theoretical and methodological tools to conceptualize and approach disastrous landscapes. In this sense, the ethnographic experiences analyzed here constitute significant contributions. The different ideas regarding the geosocial worlds of mud, mountains, rivers, rocks, and the underground are of particular significance. These ideas interestingly resonate with contemporary discussions on the vitality of matter (Bennett, 2010) and the relationships between the body and the materiality of terrain (Gordillo, 2018).

The third implication highlights the importance of interdisciplinarity. I mentioned earlier the need to explore other theories and methods in areas outside the social sciences such as geomorphology and climatology. This Dialogue is necessary in order to better understand the materiality of disaster. Yet it is also important to build bridges with other approaches to disaster in the social sciences. In Colombia, historians, sociologists, and geographers, among others, have developed studies on a variety of calamities (e.g., Lampis & Rubiano, 2012; Martínez Ardila et al., 2010; Sánchez-Calderón, 2018). This dialogue encourages anthropologists to consider other methodological paths such as archival work and spatial analysis. Furthermore, Colombian anthropologists are also developing important discussions about climate change (e.g., Camargo & Ojeda, 2017; Ulloa, 2010, 2018), which constitutes a fertile terrain for the theorization of disastrous landscapes. The Anthropology of Disasters in Colombia is still a work in progress, but we already have a diverse and empirically solid foundation. Our next step is to examine the possibilities of initiating a conversation between our ethnographies and territorial realities with the existing knowledge produced in the Americas and beyond.

Through that conversation, we will be able to discern whether those international traditions resonate with the disastrous landscapes and experiences of Colombia, or not. I hope this chapter is a contribution to that end.

The ethnographers

Mónica Cuéllar is a PhD candidate in anthropology at McGill University in Canada. Her dissertation explores the experience of "staying" in relatively abandonded rural areas of Boyacá with attention to the intimate connections between narrative, affect, and the material environment.

Lorenzo Granada is currently a master's student in anthropology at the New School in the United States. Her thesis analyzes the materiality of mud in Armero.

Marie McDonald is a PhD candidate in anthropology at the University of California, Davis. Her dissertation project studies the social lives of mudslides in Manizales. She is interested in environmental anthropology and the politics of mudslide vulnerability in Colombia.

Meghan Morris is a Postdoctoral Fellow in Law and Inequality at the American Bar Foundation and also a senior researcher at Dejusticia. She is interested in the relationships between property and conflict.

Andrés Ospina teaches at the Social Sciences School at the Universidad Pedagógica y Tecnológica de Colombia (Pedagogic and Technological University of Colombia) in Tunja. He is interested in the relationships between life and death in contexts of violence and social conflict.

Hugo Portela is a medical anthropologist who teaches at the Department of Anthropology at the University of Cauca.

Luis Alberto Suárez Guava teaches at the Anthropology Department at the Universidad de Caldas (Caldas University) in Colombia. His research agenda revolves around the understanding of the myriad theories of the world that emerge when people encounter with geological formations and phenomena.

Kristina Lyons is an Assistant Professor of Anthropology at the University of Pennsylvania. She studies the science, technologies, and politics of humans-soil relations, and is also interested in issues of socioecological justice.

References

Acero Pulgarín, S. (2010) *Encanto y temor: representaciones sociales en torno al volcán Nevado del Ruíz en el norte del Tolima*. B.A. Thesis. Bogota, Universidad Nacional de Colombia.

Baez Ullberg, S. (2017) La contribución de la antropología al estudio de crisis y desastres en América Latina. *Iberoamerican–Nordic Journal of Latin American and Caribbean Studies*. 46 (1), pp. 1–5.

Barrios, R. E. (2017) What Does Catastrophe Reveal for Whom? The Anthropology of Crises and Disasters at the Onset of the Anthropocene. *Annual Review of Anthropology*. 46, pp. 151–166.

Bennett, J. (2010) *Vibrant Matter: A Political Ecology of Things*. Durham, Duke University Press.

Bernal Gamboa, E. (2016) *Antropología en Colombia en la década de 1970. Terrenos revolucionarios y derrotas pírricas*. Bogota, Universidad Nacional de Colombia.

Buitrago Ospina, A. (2012) *Madres de crianza, levantando vida en el norte del Tolima. Estudio etnográfico de las prácticas de crianza y adopción*. B.A. Thesis. Bogota, Universidad Nacional de Colombia.

Cagüeñas, D. (2013) Mundos en la estela del desastre, esbozos para una historia de la finitud. In: Acosta, M. R. & Manrique, C. (eds.). *A la sombra de lo político, violencias institucionales y transformaciones de lo común*. Bogota, Universidad de los Andes, pp. 223–240.

Camargo, A. (2016) *Disastrous Water, Renascent Land, Politics and Agrarian Transformations in Postdisaster Colombia*. Ph.D. Thesis. Syracuse, Syracuse University.

Camargo, A. & Cortesi, L. (2019) Flooding Water and Society. *Wiley Interdisciplinary Reviews: Water*. 6 (5), n/a.

Camargo, A. & Ojeda, D. (2017) Ambivalent Desires: State Formation and Dispossession in the Face of Climate Crisis. *Political Geography*. 60, pp. 57–65.

Campos García, A., Holm-Nielsen, N., Diaz, G. C., Rubiano Vargas, D. M., Costa, P. C. R., Ramírez Cortés, F. & Dickson, E. (2011) *Analysis of Disaster Risk Management in Colombia: A Contribution to the Creation of Public Policies*. Washington, World Bank.

Cancimance, J. A. (2015) Los silencios como práctica de resistencia cotidiana, narrativas de los pobladores de El Tigre, Putumayo, que sobrevivieron al control armado del Bloque Sur de las AUC. *Boletín de Antropología*. 30 (49), pp. 137–159.

Caviedes, M. (2007) Antropología apócrifa y movimiento indígena: algunas dudas sobre el sabor propio de la antropología hecha en Colombia. *Revista Colombiana de Antropología*. 43, pp. 33–59.

Caviedes, M. (2002) Solidarios frente a colaboradores. Antropología y movimiento indígena en el cauca en las décadas de 1970 y 1980. *Revista Colombiana de Antropología*. 38, pp. 237–260.

Chaves, M. (2010) Movilidad espacial e identitaria en Putumayo. In: Chaves, M. & Del Cairo, C. (eds.). *Perspectivas antropológicas sobre la Amazonía contemporánea*. Bogota, ICANH, pp. 81–103.

Claus, C. A., Osterhoudt, S., Baker, L., Cortesi, L., Hebdon, C., Zhang, A. & Dove, M. R. (2015) Disaster, Degradation, Dystopia. In: Bryant, R. (ed.). *The International Handbook of Political Ecology*. Cheltenham and Northampton, Edward Elgar Publishing, pp. 291–304.

Cuéllar, M. (2011) Por ti me estoy consumiendo: cuerpo, despecho y brujería en el norte del Tolima. *Maguaré*. 25 (2), pp. 65–88.

Cuéllar, M. (2010) *Tragedia en clave de despecho, cuerpo, música y liminalidad en la tragedia de Armero*. B.A. Thesis. Bogota, Universidad Nacional de Colombia.

Douglas, M. & Wildavsky, A. (1982) *Risk and Culture: An Essay on the Selection of Technological and Environmental Dangers*. Berkeley, University of California Press.

Espinosa Baquero, A. (1997) Fuentes y estudios sobre desastres históricos en Colombia: retrospectiva y estado actual. In: García-Acosta, V. (ed.). *Historia y desastres en América Latina*. Lima, LA RED-CIESAS-Tercer Mundo Editores, vol. II, pp. 289–315.

Faas, A. J. (2016) Disaster Vulnerability in Anthropological Perspective. *Annals of Anthropological Practice*. 40 (1), pp. 14–27.

Findji, M. T. & Bonilla, V. D. (1995) ¿El otro, el mismo? Tragedias, cultura y lucha entre los Paeces. *Desastres y Sociedad*. 4 (3), pp. 46–59.

García-Acosta, V. (2005) El riesgo como construcción social y la construcción social del riesgo. *Desacatos: Revista de Antropología Social*. 19, pp. 11–24.

García-Acosta, V. (2004) La perspectiva histórica en la Antropología del riesgo y del desastre. Acercamientos metodológicos. *Relaciones. Estudios de historia y sociedad.* XXV (97), pp. 123–142.

García-Acosta, V. (2002) Historical disaster research. In: Hoffman, S. M. & Oliver-Smith, A. (eds.). *Catastrophe & Culture: The Anthropology of Disaster.* Santa Fe, Nuevo Mexico, School of American Research, pp. 49–65.

GFDRR. (2018) *Colombia.* Available from www.gfdrr.org/en/colombia. Accessed April 6th 2019.

Girot, C. & Imhof, D. (2016) Introduction. In: Girot, C. & Imhof, D. (eds.). *Thinking the Contemporary Landscape.* New York, Princeton Architectural Press, pp. 7–14.

Gordillo, G. (2018) Terrain as Insurgent Weapon: An Affective Geometry of Warfare in the Mountains of Afghanistan. *Political Geography.* 64, pp. 53–62.

Gordillo, G. (2014) *Rubble: The Afterlife of Destruction.* Durham, Duke University Press.

Gow, D. (2008) *Countering Development: Indigenous Modernity and the Moral Imagination.* Durham, Duke University Press.

Granada, L. (2016) Etnografía en fragmentos: escombros, ruinas y ausencias en el Valle de Armero. *Ecuador Debate.* 99, pp. 79–101.

Henare, A. J. M., Holbraad, M. & Wastell, S. (2007) Introduction, Thinking Through Things. In: Henare, A. J. M., Holbraad, M. & Wastell, S. (eds.). *Thinking Through Things: Theorizing Artefacts Ethnographically.* London, Routledge, pp. 1–30.

Henry, D. (2005) Anthropological Contributions to the Study of Disasters. In: McEntire, D. & Blanchard, W. (eds.). *Disciplines, Disasters and Emergency Management: The Convergence and Divergence of Concepts, Issues and Trends from the Research Literature.* Emmitsburg, Federal Emergency Management Agency.

Hirsch, E. (1995) Landscape: Between Place and Space. In: Hirsch, E. & O'Hanlon, M. (eds.). *The Anthropology of Landscape: Perspectives on Place and Space.* Oxford and New York, Clarendon Press/Oxford University Press, pp. 1–30.

Horisberger, N. (2018) Of Salt and Drought – What Methods for Ethnographic Research in Fluid Places? In: Krause, F. (ed.). *Delta Methods: Reflections on Researching Hydrosocial Lifeworlds.* Cologne, University of Cologne. Available from https://kups.ub.uni-koeln.de/8961/1/KAE-07_Krause2018_Delta%20Methods.pdf. Accessed April 6th 2019.

Ingold, T. (1993) The Temporality of the Landscape. *World Archaeology.* 25 (2), pp. 152–174.

Ingold, T. & Vergunst, J. L. (eds.). (2008) *Ways of Walking: Ethnography and Practice on Foot.* Aldershot, Ashgate.

Ishizawa, O. A. (2018) *The Disaster Risk Management Challenge for Small Cities.* Available from https://blogs.worldbank.org/latinamerica/disaster-risk-management-challenge-small-cities. Accessed April 6th 2019.

Jimeno, M. (2018) Anthropology in Colombia. In: Callan, H. (ed.). *The International Encyclopedia of Anthropology.* New York, John Wiley & Sons, Ltd. Available from https://doi.org/10.1002/9781118924396.wbiea1378. Accessed April 6th 2019.

Kusenbach, M. (2003) Street Phenomenology: The Go-Along as Ethnographic Research Tool. *Ethnography.* 4 (3), pp. 455–485.

Lampis, A. & Rubiano, L. (2012) ¡Y siguen culpando a la lluvia! Vulnerabilidad ambiental y social en el Altos de la Estancia, Bogotá, Colombia. In: Briones, F. (coord.). *Perspectivas de investigación y acción frente al cambio climático en Latinoamérica.* Mérida, LA RED, pp. 177–220.

Lyons, K. (2018) Los ríos tienen memoria: la (im)posibilidad de las inundaciones y historias de (de) y (re) construcción urbana en el piedemonte amazónico. *A la orilla del río*. Available from http://alaorilladelrio.com/2018/05/30/los-rios-tienen-memoria-la-imposibilidad-de-las-inundaciones-e-historias-de-de-y-re-construccion-urbana-en-el-piedemonte-amazonico/. Accessed April 6th 2019.

Martínez Ardila, N. J., Corrales Cobos, J. J. & Sánchez Calderón, V. (2010) Relación de los deslizamientos y la dinámica climática en Colombia. In: Guerrero Barrios, V., Garzón Vargas, V. & Carmona Cuervo, J. (eds.). *Experiencias en el uso y aplicación de tecnologías satelitales para observación de la tierra*. Bogota, IDEAM, pp. 79–97.

Meneses Lucumí, L. E. (2000) *De la montaña al valle, tradición y cambio en el resguardo de Caloto*. B.A. Thesis. Popayán, Universidad del Cauca.

O'Keefe, P., Westgate, K. & Wisner, B. (1976) Taking the Naturalness Out of Natural Disasters. *Nature*. 260, pp. 566–567.

Oliver-Smith, A. (2002) Theorizing Disasters: Nature, Power, and Culture. In: Hoffman, S. M. & Oliver-Smith, A. (eds.). *Catastrophe & Culture: The Anthropology of Disaster*. Santa Fe, Nuevo México, School of American Research, pp. 23–48.

Oliver-Smith, A. (1999) "What Is a Disaster?" Anthropological Perspectives on a Persistent Question. In: Oliver-Smith, A. & Hoffman, S. M. (eds.). *The Angry Earth: Disaster in Anthropological Perspective*. New York and London, Routledge, pp. 17–34.

Oliver-Smith, A. & Hoffman, S. M. (2002) Introduction: Why Anthropologists Should Study Disasters. In: Hoffman, S. M. & Oliver-Smith, A. (eds.). *Catastrophe & Culture: The Anthropology of Disaster*. Santa Fe, Nuevo México, School of American Research, pp. 3–22.

Ospina, A. (2013) El sacrilegio sagrado, narrativa, muerte y ritual en las tragedias de Armero. *Revista Colombiana de Antropología*. 49 (1), pp. 177–198.

Pineda, R. (2004) La escuela de antropología colombiana: notas sobre la enseñanza de la antropología. *Maguaré*. 18, pp. 59–85.

Quiceno Montoya, C. N. (2018) *Reconstrucción del tejido social después de un desastre, mujeres y territorio en la avalancha del río Páez 1994. Una mirada feminista de la realidad*. PhD Thesis. Madrid, Universidad Complutense de Madrid.

Ramírez, M. C. (2001) *Entre el Estado y la guerrilla, identidad y ciudadanía en el movimiento de los campesinos cocaleros del Putumayo*. Bogota, ICANH, Colciencias.

Ramírez Elizalde, L. A. (2013) *¿Irse, quedarse o llevar el territorio a cuestas? El proceso de reorganización nasa después del terremoto*. M.A. Thesis. Bogota, Universidad de los Andes.

Rappaport, J. & Gow, D. (1997) Cambio dirigido, movimiento indígena y estereotipos del indio. El Estado colombiano y la reubicación de los Nasa. In: Uribe, M. V. & Restrepo, E. (eds.). *Antropología de la modernidad*. Bogota, ICAN, pp. 361–399.

Restrepo, E. (2016) La antropología en Colombia en el nuevo milenio. In: Tocancipá, J. (ed.). *Antropologías en Colombia: tendencias y debates*. Popayán, Universidad del Cauca, pp. 63–84.

Sánchez, A. (2011) *Después de la inundación*. Cartagena, Centro de Estudios Económicos Regionales.

Sánchez-Calderón, V. (2018) Agua y desigualdades socio-ecológicas en Bogotá a mediados del siglo XX. El caso del río Tunjuelo y sus barrios ribereños. In: Ulloa, A. & Romero-Toledo, H. (eds.). *Agua y disputas territoriales en Chile y Colombia*. Bogota, Universidad Nacional de Colombia, pp. 391–427.

Sarabia Gómez, A. M., Cifuentes Avendaño, H. G. & Robertson, K. (2010) Análisis histórico de los sismos ocurridos en 1785 y en 1917 en el centro de Colombia. *Cuadernos de Geografía, Revista Colombiana de Geografía*. 19, pp. 153–162.

Serna Quintana, C. A. (2011) La naturaleza social de los desastres asociados a inundaciones y deslizamientos en Medellín 1930–1990. *Historia Crítica*. 43, pp. 198–223.

Stewart, P. & Strathern, A. (2003) Introduction. In: Stewart, P. & Strathern, A. (eds.). *Landscape, Memory and History: Anthropological Perspectives*. London and Sterling, Pluto, pp. 1–15.

Stoler, A. L. (ed.). (2013) *Imperial Debris: On Ruins and Ruination*. Durham, Duke University Press.

Suárez Guava, L. A. (2009) Lluvia de flores, cosecha de huesos: guacas, brujería e intercambio con los muertos en la tragedia de Armero. *Maguaré*. 23, pp. 371–416.

Suárez Guava, L. A. (2008) Juan Díaz engañado por la riqueza: un artífice de la fortuna y la tragedia en un mundo colonial. *Maguaré*. 22, pp. 223–298.

Tabares, E. & Meneses, L. E. (2016) *Historia de la antropología en el Cauca*. Popayán, Universidad del Cauca.

Telesur. (2017) Confirman 321 fallecidos por avalancha en Mocoa, Colombia. *Telesur*. Available from www.telesurtv.net/news/Confirman-321-fallecidos-por-avalancha-en-Mocoa-Colombia-20170414-0047.html. Accessed April 6th 2019.

Therrien, M. (1995) Terremotos, movimientos sociales y patrones de comportamiento cultural: arqueología en la cubierta de la Catedral Primada de Bogotá. *Revista Colombiana de Antropología*. 32, pp. 148–183.

Tilley, C. Y. & Cameron-Daum, K. (2017) *Anthropology of Landscape: The Extraordinary in the Ordinary*. London, UCL Press.

Tocancipá, J. (ed.). (2016) *Antropologías en Colombia: tendencias y debates*. Popayán, Universidad del Cauca.

Tsing, A. (2014) Blasted Landscapes and the Gentle Arts of Mushroom Picking. In: Kirksey, E. (ed.). *The Multispecies Salón*. Durham and London, Duke University Press, pp. 87–109.

Ulloa, A. (2018) Reconfiguring Climate Change Adaptation Policy, Indigenous Peoples' Strategies and Policies for Managing Environmental Transformations in Colombia. In: Klepp, S. & Chavez-Rodríguez, L. (eds.). *A Critical Approach to Climate Change Adaptation: Discourses, Policies and Practices*. London, Routledge Advances in Climate Change Research, pp. 222–237.

Ulloa, A. (2010) Geopolíticas del cambio climático. *Anthropos*. 227, pp. 133–146.

UNGRD. (2018) *Atlas de riesgo de Colombia: revelando los desastres latentes*. Bogota, Unidad Nacional para la Gestión del Riesgo de Desastres.

Uribe de Zuluaga, E. (2009) Final del mundo en Armero. *Maguaré*. 23, pp. 513–532.

Wilches-Chaux, G. (1997) Prólogo. In: Gómez, H. & Ruíz, C. (eds.). *Los Paeces: gente territorio, metáfora que perdura*. Popayán, Universidad del Cauca.

World Bank. (2016) *Colombia, Policy Strategy for Public Financial Management for Natural Disaster Risk*. Washington, World Bank.

World Bank. (2013) *Colombia: los desastres del futuro afectarán principalmente a las ciudades*. Available from www.bancomundial.org/es/news/feature/2013/02/14/colombia-future-disasters-will-mainly-affect-cities. Accessed April 6th 2019.

Zeiderman, A. (2012) On Shaky Ground: The Making of Risk in Bogota. *Environment and Planning A: Economy and Space*. 44 (7), pp. 1570–1588.

5 Anthropologies of Disasters in Ecuador

Connections and apertures

A.J. Faas

Introduction: hazards, ecology, and engagements[1]

Ecuador's varied topography – broadly divided into coastal, highland, and Amazonian regions – is notable for exceptionally rich biodiversity, numerous microclimatic zones, and a prevalence of environmental hazards.[2] Each of the broad topographic regions hosts particular arrays of hazards, some of which, such as landslides, occur where these regions meet. Common highlands hazards include earthquakes, landslides, drought, and volcanoes – 29 on the mainland, another 15 in the Galapagos Islands. Some volcanoes are not known to have been active for more than a thousand years, but several have registered eruptions over the past 200+/- years. Among the most active are Cotopaxi, Guagua Pichincha, Reventador, Tungurahua, and Sangay, the latter considered the most continuously active volcano in Latin America. The coastal region is the epicenter of El Niño Southern Oscillation activity in the Western Hemisphere and other coastal hazards include flooding, landslides, and earthquakes and tsunamis – the Nazca tectonic plate diving beneath the Pacific Coast is known to skip in convulsive jolts. Principal hazards in the Amazonian region are environmental degradation resulting from oil extraction and deforestation. Seismic and volcanic events and processes often feature prominently in oral histories of indigenous peoples and rural *campesinos*. El Niño activity is particularly salient in the archaeological record and has, since the turn of the 21st century, reemerged as an important focal point in disaster research in Ecuador due both to its calamitous impacts and its position at the intersection of climate science and climate change discourses.

Ecuadorian history is replete with stories of disasters, many of which have not been substantively examined by anthropologists or other social scientists. The 20th century saw a proliferation of technological hazards, especially those associated with oil extraction in the Amazon and deforestation along the coast. Archaeological evidence points to the centrality of hazards in the development of early civilizations in Ecuador, Spanish chroniclers actively monitored volcanic activity, and mountain expeditions, major landslides, earthquakes, and eruptions have featured prominently in Ecuadorian affairs from the 19th through the 21st century. Yet there has been surprisingly little anthropological attention paid to the topic of disasters in Ecuador. Most studies focused on the topic have come from the global north and only within the past two decades. However, a closer reading of anthropological and other social science studies in Ecuador finds that the social, cultural, and environmental impacts of geophysical, hydrometeorological, and technological phenomena are surfaced in ethnographic and other critical works as well as in local memories and narratives of landscape.

Writing shortly before his death in 1999, influential Ecuadorian[3] economist Germánico Salgado Peñaherrera (1999) was concerned to articulate a vision of Ecuador's future. The 1997–1998 El Niño was at the time the most powerful in recorded history and Ecuador sustained major impacts – hundreds were killed in landslides and thousands more sustained injuries and property damage, with total damages exceeding US$2.8 billion. Half the country's transportation infrastructure was left in shambles and a large portion of its production lost. Salgado predicted that recovery efforts would take many years at a cost that would overshadow Ecuador's economy. The loss of production damaged Ecuador's export economy at a time when it was most crucial. Moreover, as El Niño wreaked havoc, an earthquake struck the part of the country hit hardest by the climatic phenomenon. At the very same time, Quito was in "yellow" (a nonemergency, preventative) alert in anticipation of an eruption from Guagua Pichincha, whose peak looms large over the center of the city. Salgado described Ecuador as approaching an inflection point – as a country moving away from several years of upheaval and desperation – and spoke of historic changes in perspective, institutions, and overall thought (Salgado, 1999: p. 69). He saw two historic forces bringing change in the country: direct confrontations with Ecuador's past and the political and economic crises that characterized the 1990s. In this short essay, Salgado referred to Ecuador's environmental state as beset by multiple crises – a combination of past calamities for which the Ecuadorian people were responsible, external forces the Ecuadorian people could do nothing about, and crises resulting from hazards so prominent in the country's geography (Salgado, 1999: p. 68).

Though it came to be much more clearly articulated in the 1990s (e.g., García-Acosta, 1993; Maskrey, 1993; Lavell, 1993; Blaikie et al., 1994; Oliver-Smith & Hoffman, 1999), engagements with disasters in anthropology and geography have been predicated on a political ecological framework that rejects the notion of disasters as natural, focuses on the historical production and distribution of risk by human agents, and places power, postcoloniality, vulnerability, everyday structural violence, and subalternity at the center of analysis (Sun & Faas, 2018; Faas, 2016). Research in the 21st century typically proceeds with much of this firmly established and therefore largely implied, as investigators critically evaluate competing narratives of disaster, risk perception, expert and local knowledges, memory and memorialization, risk reduction, recovery, response, and overall a critical attention to the role of culture and politics in these processes and encounters (Baez Ullberg, 2017; Kulstad-González & Faas, 2016). Most scholars in Latin America enter global conversations about disasters and vulnerability through the work of Virginia García-Acosta (2005), Allan Lavell (1993), and Andrew Maskrey (1993), all of whom have been central to advancing critical theories of disaster in Latin America for decades (Faas & Barrios, 2015). Yet, since many of these developments also reflect other trends in the discipline – problematizing human-environmental relations; postcolonial critique and the discursive and dialectical relationships between past, present, and future; decolonizing methodologies; and the ontological turn – it is, at times, possible to critically engage disasters without explicitly referencing the core body of disaster anthropology in the political ecological vein.

In this short review of anthropological (and related) engagements with disasters in Ecuador, I call attention to research that is explicitly and implicitly situated in these ongoing conversations and a few instances where there are notable gaps. In light of Ecuadorian confrontations with enduring legacies of colonialism, the economic crises of the final decades of the 20th century, neoliberal reforms and indigenous uprisings in the 1990s, and the transformations of the Ecuadorian state in the first decades of the 21st century, I review scholarship on disasters with attention to historical processes and tensions, political ecology, and emergent conceptual frameworks. This chapter reviews important formal treatments of disaster production, prevention, response, recovery, and resettlement, and adaptation to chronic hazards in anthropological literature, with attention to archaeological investigations, ethnohistorical research, medical anthropology, social network analyses, and critical studies of the intersections of Ecuadorian traditions, governance, and expert knowledge in disaster recovery and resettlement. This is complemented by a critical analysis of how other anthropologies point to important hazard- and

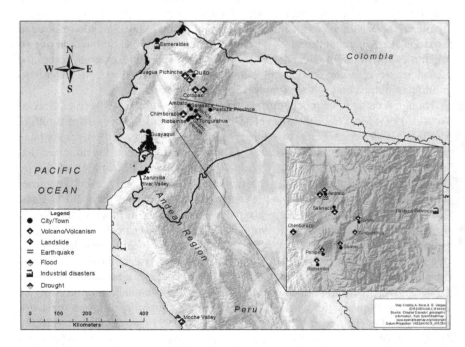

Map 5.1 Map Ecuador. Case studies and main areas mentioned

disaster-related dynamics in Ecuadorian society and culture and indigenous and rural *campesino* narratives reveal the historically intimate and adaptive relationships between people, memory, landscape, place, and the hazards, both environmental and technological, which affect these relationships.

Anthropological studies of pre-Columbian hazards and disasters

Ecuador's highest mountain, the stratovolcano Chimborazo, is one of many long revered by indigenous peoples of Ecuador; and archaeologists have identified remnants of what may have been pre-Columbian sites of offerings to the volcano (Haro, 1977: pp. 103–104). Nearby stratovolcano Tungurahua is commonly recognized as the female counterpart to the masculine Chimborazo. *Huacas* (or *wak'a*) are beings that inhabit (or simply *are*) what in the global north are commonly thought of as environmental features – especially mountains and volcanoes; a sort of "natureculture" (Haraway, 2016 [2003]) defying the socially constructed neat divide between domains of "nature", "supernatural", and "social". Silvio Luis Haro Alvear, who would go on to become a Monsignor in the Archdiocese of Riobamba, studied history at the Catholic University of Lyon and wrote prolifically about pre-Columbian civilization in Ecuador, chiefly the Puruhá (or Puruháes), who have occupied the territories known today as northern Chimborazo, Bolivar, and Tungurahua provinces since before Incan expansion into the region.[4] Haro wrote of the early development of Puruhá society as aided by their settlement on a mantle of soils and sediments rich in volcanic ash (Haro, 1977: p. 15). In fact, the Puruhá speak of their ancestors as emerging from the snowcapped Chimborazo volcano (González Suárez, 1890: p. 107). Haro (1977: pp. 102–110) described the pre-Columbian reverence for volcanoes as sacred beings and throughout his several volumes on pre-Columbian cultures of Ecuador (see especially Haro, 1973), he paid significant attention to volcanic eruptions and earthquakes as influencing the development of these societies. For instance, he wrote of the pre-Incaic culture of Guano being destroyed by an eruption (Haro, 1977: p. 49), though he generally noted such events without attempting to theorize or investigate them. He did, however, discuss the enduring significance of volcanoes and earthquakes in the practices of contemporary indigenous peoples of Chimborazo (Haro, 1977: p. 93).

Contemporary archaeological investigations of disaster in Ecuador principally focus on interpreting pre-Columbian material culture and the ways in which this evidence informs important conclusions regarding patterns of production, exchange, consumption, settlement, adaptation to chronic hazards, and the emergence of inequality. Mexican archaeologist Linda Manzanilla (1997) studied how the phenomenon of El Niño factored in various epochs of the pre-Inca history of the Andes in a comparative study that included cases from Mesoamerica and Egypt. El Niño originates along the coasts of the Pacific Ocean and causes a combination of devastating droughts and catastrophic rains and landslides in the Andes. From the Formative Horizon in the Moche (Cupisnique) Valley on what is now the northern coast of Peru (just south of the contemporary Ecuadorian border), Manzanilla found evidence of strong El Niño events with consequent changes in subsistence and settlement patterns. Domestic middens dated between 1300 and 300 BCE revealed deposits of mollusks, crustaceans, and tropical fish associated with hot water incursions revealing changes in biological chains of the Pacific currents and adaptations in shore fishing technologies (Manzanilla, 1997: p. 50). It was likely droughts triggered by El Niño, such as those that occurred around 700 and 1100 CE, which caused profound changes and the subsequent collapse of the Tiwanak system, inducing severe tensions in agricultural and social systems. The frequent incidence of climatic disturbances associated with El Niño brought about demographic rearrangement, changes in settlement patterns and foodways, architectural reconstruction, the application of flood control, and agricultural intensification technologies, and significant ideological changes (Manzanilla, 1997: p. 51). The ritual exchange of the bivalve *spondylus princeps* (*mullu* in Quichua/Quechua), which is closely related to the warm currents of El Niño, but variation in their coastal population also likely served as a climatic indicator to determine cycles of rain and drought. The priests of the oracles came to use the *mullu* and observatories as instruments of prediction.

American archaeologist Sarah Taylor (2015) studied the prehistoric Zarumilla River Valley along the contemporary Peru-Ecuador border. She uncovered archaeological evidence of the emergence of inequality as part of changes in a suite of exchange practices and settlement patterns in response to severe El Niño events. Her findings indicate that broad cultural adaptations – in the form of exclusive exchange relations and wealth accumulation – to chronic and acute hazards increased vulnerability for some, while reducing it for the emergent elite. These findings have implications for how we think of such concepts as resilience today, as we search for ways to foster sustainable development that is both robust and inclusive. Taylor's archaeological exploration reveals emergent and coconstituting relationships between political systems, agricultural practices, and the environment's material agency in pre-Columbian communities and exchange networks.

These archaeological findings resonate with the present in several ways. In these cases, as often in the present, disasters not only entail hazards but social disarticulations, with groups and exchange networks fragmented and factionalized; and disaster inspires solidarity and mutual aid at times, and opportunism and the production of inequality at others. Yet, we can distinguish indigenous adaptations from subsequent colonial and modernist responses. For one, pre-Columbian modes of production involved direct engagement with their environments, which not only meant a greater sensitivity to environmental changes with specific implications for subsistence, but also a complex and deeply rooted semiotic, ritual, and religious connection to the landscape and environment (Boelens, 2015: pp. 82–88; Corr, 2010). Moreover, many pre-Incan societies had a more mobile relationship with the environment, moving seasonally between territories and subsistence niches (e.g., coast and highland). Incan and subsequently colonial and Republican colonization of smaller-scale societies frequently entailed the production of fixed (re)settlements and the enclosure and reduction of territory to "property", increasing vulnerability as a result of constrained capacity to navigate hazards, increased exposure to technological hazards, and racial, political, and economic marginalization.

Disasters in colonial and 19th-century Republican eras

Disaster processes of the colonial and 19th-century Republican eras have yet to be substantively addressed by anthropologists. During the colonial period, Spanish settlers and administrators kept journals documenting major volcanic activity – making note of eruptions of Cotopaxi and Tungurahua in 1534 (Rachowiecki, 2008: p. 148). Earthquakes in 1797 destroyed Chimborazo's capital city of Ambato, during which noted *cacique* (indigenous leader) Hernando of Cagi (or "Caguají") disappeared; the Chambo River (which runs north from Chimborazo Province alongside the volcano Tungurahua into Tungurahua Province) was dammed, and the population devastated; and the parish church of Guano, La Asunción, was destroyed (Haro, 1977: pp. 53–56). Ibarra's massive 1868 earthquake, the eruptions of Tungurahua in 1886, and the Great Fire of Guayaquil in 1896 were just some of the disasters that marked the end of the 19th century in Ecuador. Though there were many press accounts of the Guayaquil fire, the absence of substantive critical social science engagements with these disasters leaves important gaps in our understanding of their place in Ecuadorian culture and society.

One interesting exception is to be found in Argentinian anthropologist Blanca Muratorio's (1998) study of statecraft in the Amazon during the Republican period of the 1850s. Though this work does not deal with disaster per se, Muratorio found that the region was subjected to a sort of malign neglect by the state in the 19th century. As a result of state abandonment, indigenous peoples lacked legal protection while private commercial ventures enjoyed free reign to pursue their own interests. This would have repercussions well into the 20th and 21st centuries, which I discuss in the following section.

Travel writer Rob Rachowiecki (2008) has written of 19th-century travelers' and mountaineers' fascination with the volcanoes of Ecuador. This can, perhaps, be seen as an extension of European Enlightenment and colonial imaginaries of conquering nature and the spaces of tropical "other worlds" (Tsing, 2005: pp. 131–146; Bankoff, 2001). French geologist Sebastian Wisse recorded 267 major explosions of Sangay in one hour during the first recorded expedition to the volcano in 1849 (Rachowiecki, 2008: p. 150). A German-Ecuadorian team reached the summits of Cotopaxi and Tungurahua in 1873. The 26 June 1877 eruption of Cotopaxi rid the northeast slopes of ice and snow for a subsequent ascent. And famous English mountaineer Edward Whymper subsequently camped for a night by Cotopaxi's crater in 1880.

Canadian anthropologist Kim Clark (2002, 1998) studied railway construction as a crucial element in Ecuadorian nation-building and modernity during the liberal period from the mid-1890s to 1930, when coastal agro-export elites rose to political dominance over the long-ascendant conservative highland landowning class. Coastal liberals were anxious to integrate the coastal and highland economies to open up new import markets and the railway was central to this unifying – and modernizing – vision. Landslides previously prevented interregional travel for several months each year and travel during optimal conditions by mule could take up to two weeks (Clark, 2002). This Ecuadorian modernization project therefore in many ways entailed a direct confrontation with environmental hazards. After independence, highland *hacendados* (individuals commanding large estates and populations of indigenous peons) and coastal plantation capitalists continued to exact labor tribute via *minga* work parties (described in greater detail later) and, while indigenous peoples and *campesinos* were popularly derided as backward and lazy, *minga* labor gangs plainly underwrote the modernization of Ecuadorian state infrastructure and economy, as they built the road, rail, and telegraph systems that united national territories (Larson, 2008: pp. 580–606). These processes are examples of the historical intersection of racial, political, infrastructural, and economic developments with hazards in Republican Ecuador at the dawn of the 20th century.

Technological hazards and impacts in the 20th and 21st centuries

Ecuador does not have any nuclear reactor facilities and radioactive waste production is minimal. Industry is principally concentrated in the dense cities of Guayaquil and Quito and most is relatively small scale and therefore not associated with sizeable toxic releases. But it is worth noting that urban anthropology – a field in which studies of technological hazards and environmental justice have flourished in other national contexts – remained relatively inchoate in Ecuador prior to the 21st century (Moreno Yánez, 1992). Today, however, urban anthropology is an active field of research in Ecuador,[5] but studies of the impacts of technological hazards are focused on Amazonian and coastal regions far removed from Ecuador's urban centers.

The predominant technological hazard comes from oil extraction in the Amazonian region, which is also the country's principal export and largest contributor to the national GDP and state revenue. Several studies since the 1990s have pointed to the disastrous impacts associated with oil extraction in the Ecuadorian Amazon (e.g., Kimerling, 1993). In an attempt to identify statistical patterns in the relationship between oil extraction, social structure, and quality of life in the Amazon, Ecuadorian anthropologist Teodoro Bustamante and socioenvironmental studies student Maria Cristina Jarrín (2005) drew on data from the Ecuadorian census (National Institute of Statistics and Census, INEC) and the Socioenvironmental Observatory of the Facultad Latinoamericana de Ciencias Sociales (Latin American Social Sciences Institute, FLACSO-Ecuador). In a finding that might seem inconsistent with other studies on this topic, they found that the presence of oil wells was a distant second to infrastructure variables as a correlate of social indicators of inequality and quality of life (including health, housing, and education). However, when taken in the context of other studies, this is rather consistent with a core argument made by disaster researchers around the world for decades: it is not the hazard, but rather the intersection of a hazard with a suite of human practices – development, environmental justice, postcolonial relationalities – and human populations that produces disaster (Sun & Faas, 2018). Thus, it is helpful to read Bustamante and Jarrín's finding about infrastructure as an indicator of the articulation of certain spaces of oil extraction with the production of the state and the development of institutions and infrastructure; this interpretation would be consistent with earlier work by Muratorio (1998, see earlier) and contemporary anthropological approaches to studying the state in Ecuador (e.g., Prieto, 2015; Krupa & Nugent, 2015). Muratorio's work well encapsulates an important current in the historical production of vulnerability in the Amazon, while recent phenomenological approaches to studying the state in the Andes (and beyond) beckon us to consider its incoherencies and inseparability from varieties of cooperation and competition among racial, class, and ethnic interests. These trends are well represented in several important ethnographic studies of extraction and indigenous livelihoods and politics in the Amazon.

Though neither engages the core literatures on the Anthropology of Disasters, American anthropologists Suzanna Sawyer and Robert Wasserstrom have produced some of the most compelling anthropological studies of the intersection of development, ecology, indigenous human rights and resistance movements, and subjectivity in the contexts of the proliferation of technological hazards and intrusions associated with petroleum extraction in the Amazon. Wasserstrom (2016) has studied indigenous resistance to extractive incursions of the petroleum industry in the Amazon. Though the Waorani are often popularly imagined as isolated, uncontacted, violent, and warlike, Wasserstrom points out that they have been in continuous contact with outsiders since the late 1900s. Violent conflicts, however, had historically resulted from extractive incursions and agricultural settlement in their territories, which escalated again in the 1980s and 1990s with renewed oil development in the region (Wasserstrom, 2016: p. 500).

The discovery of commercially viable petroleum reserves in the Amazon by Texaco in 1967 led to these reserves being brought into production in 1972 and Ecuador joining OPEC in the following year (Sawyer, 2004: p. 11). State oil revenues have been central to the expansion of infrastructure and social services from the 1970s through the present, though extensive borrowing against future yields, along with price drops and rising lending rates, led to a major economic crisis in the early 1980s. Suzana Sawyer's (2004, 2008) ethnographies of indigenous Amazonians' resistance to largely unfettered oil extraction revealed disastrous outcomes and associated health and environmental impacts. But her work goes well beyond this to investigate the political ecology of extraction, bringing into relief the historical and collective claims of indigenous peoples as they confronted the neoliberalizing agendas of the Ecuadorian state and the racial and colonialist discourses underwriting them in the 1990s. The Trans-Andean pipeline leaked nearly 17 million gallons of crude oil in the Amazonian waters between 1972 and 1990 and spillage from secondary lines is said to account for comparable volumes (Sawyer, 2004: p. 101). In a case of convergent catastrophe (see Moseley, 1999), Texaco executives pointed to the 1987 earthquake as causing pipeline fractures accounting for a significant portion of spill volume. Toxic components of crude oil have been associated with cancers, reproductive abnormalities, nerve damage, and skin disease, and multiple toxic pollutants are also associated with the extraction and production processes (Sawyer, 2004: p. 102). In addition to processing facilities, pumping stations, and a refinery, Texaco's operations resulted in vast seismic grids, more than 300 wells, and over 600 open waste pits littering the Amazonian landscape (Sawyer, 2008: p. 322). The result, as conveyed by one of Sawyer's interlocutors, was "Our children kept on getting sicker . . . as the toxins and contamination got worse. The doctor told my neighbor that her five-year-old son had leukemia. That's when we started to organize ourselves" (Sawyer, 2008: p. 325).

Sawyer's ethnography focuses primarily on the strategies of the Organization of Indigenous Peoples of Pastaza (OPIP) to resist environmental devastation and the commodification of indigenous territories in the Amazon. She captures the many encounters and conflicts revealing enduring racial and class hierarchies in Ecuadorian society, with attention to white neoliberal agendas focusing on capitalist production, developing and exporting resources, and deregulation. One core issue is that, while OPIP argued for "historical" claims to "collective territories", the state and dominant economic interests claim the right to allocate "property" rights to individuals and corporate interests. The mission to civilize indigenous peoples and territories remains central to dominant Amazonian imaginaries and dominant discourse focuses on ethnicity and ethnic difference, whereas indigenous peoples speak of historical claims, precolonial territories, and legacies of injustice and invasion. While indigenous activists are concerned with "moral-cosmological and political-economic complexes that shape identity and social relations", the Ecuadorian state was preoccupied with "rationally calculated scientific blocks . . . codified and incarcerated through lines on a map" (Sawyer, 2004: p. 83).

Tungurahua in the 21st century: multidisciplinary applied social science and critical studies

Disasters do not result from isolated events but are instead products of the collision of historical processes – development, colonialism, quotidian structural violence – and hazards; at times, multiple such crises converge (Moseley, 1999). And, at the turn of the 21st century, Ecuador was confronting a particularly vexing convergence. The massive El Niño of 1997–1998 came during a fall in petroleum prices, followed by hyperinflation, a banking crisis, and revelations of rampant corruption in provincial and federal governments. Within a year, the Ecuadorian government would move to abandon the national currency, the *sucre*, and adopt the American dollar. In October 1999, stratovolcano Tungurahua erupted, displacing tens of thousands and devastating homes and agricultural and livestock production throughout the rural regions of Chimborazo and Tungurahua Provinces. Months later, President Mahuad was driven to resign by indigenous protest and a military revolt.

Several anthropological studies of disaster in Ecuador have taken shape around Tungurahua since it began its renewed eruptive phase in 1999. Before it came roaring back to life in October of that year, Tungurahua was last active from 1916 to 1925, when the volcano generated ashfalls and intense pyroclastic flows and then returned to slumber for nearly 80 years. Occasionally, someone from the northern villages of Cantón Penipe will recall hearing stories of the 1914–1918 eruptions from their parents or grandparents who remembered days of darkness as clouds of ash blackened the sky and tremors shook the earth as pyroclastic flows ran down the mountain. In this section, I review some of the applied anthropological projects and critical studies that took shape around the 1999 and 2006 eruptions of Tungurahua. I begin with attention to medical anthropological projects and follow this with studies of social support and critical analyses of the intersection of expert interventions and local cultural practices in post-disaster recovery and resettlement.

Applied medical anthropology: health impacts and social support

Following the 1999 eruptions of Tungurahua, American medical anthropologist Linda Whiteford and British-American geographer Graham Tobin of the University of South Florida began a decades-long collaborative project with American anthropologists Art Murphy and Eric Jones from the University of North Carolina at Greensboro and Hugo Yepes from the Ecuadorian Instituto Geofísico,[6] to study impacts and recovery. It began with a focus on risk perception and medical anthropology, but later developed into several multidisciplinary collaborations focused on social networks and social support, the political ecology of disasters, the intersections of highland cultural practice and humanitarian interventions and, ultimately, disaster-induced resettlements after the devastating eruptions of 2006. Although Tungurahua has been scientifically monitored by the Instituto Geofísico since 1989, no official information on risks posed by the volcano was actively disseminated to the public before technical alerts in the month prior to the 1999 eruptions. At the time, as in most Latin American states, disaster relief was primarily handled by the Civil Defense forces, with higher level personnel being primarily retired military. Mexican geographer Jesus Manuel Macías and sociologist Benigno Aguirre (2006: p. 53) studied the 1999 evacuation of Baños where state authorities shut down electricity, water flow, and telephone service to hasten evacuation and discourage premature return. During the prolonged evacuation, tensions ran so high that residents began aggressively confronting state police in order to return to their homes, in some instances actually sequestering military officers in order to negotiate their surrender (Macías & Aguirre, 2006: p. 53). Whiteford and Tobin (2004; Tobin & Whiteford, 2002) tell of the complexities of evacuation and response to the eruptions and the diverse and contrasting perceptions among business owners, Defensa Civil, technological experts at the Instituto Geofísico, and residents.

Following the 1999 eruptions, American geography student Lucille Lane worked with Tobin and Whiteford (see Lane et al., 2003) to study what urban evacuees faced in deciding to return to what the Ecuadorian government deemed a high-risk area. In the small city of Baños in the northern shadow of the volcano, almost all residents were directly or indirectly involved in the tourist industry and perceived their livelihoods as endangered if they could not return in time for the tourist season. The team found that economic livelihood concerns mitigated "fear and worry about the volcano hazard" (Lane et al., 2003: p. 28). Many people placed pressure on officials to downgrade the risk category to permit Baños residents to return regardless of the expert risk pronouncements of the Instituto Geofísico. These events transpired during the worst throes of the late 1990s economic crisis, which meant that economic anxieties were particularly pronounced among evacuees, who were not simply concerned with returning home, but with securing their precarious livelihoods.

Penipe is a rural canton whose northern limits extend from the southern to the western flanks of the volcano. In 2004, American medical anthropologist Juan Luque came to Penipe to study beliefs and barriers related to maternal healthcare-seeking for young children with respiratory illnesses, with attention to acute respiratory infections (ARIs) potentially resulting from contact with volcanic ash (Luque, 2007: p. 282). Many common health problems in Ecuador are extensions of poverty and marginalization and, at the time of the study, 70 percent of children in Ecuador were living in poverty and 15 percent under the age of five were affected by malnutrition. Luque and colleagues (2008) found that caregivers were "unaware of the seriousness of ARIs due to lack of health education and health-seeking behavior", which included using home remedies of herbal teas, and seeking local traditional healers before seeking medical assistance. Although caregivers mentioned the cost of care and medicine as a barrier to seeking formal healthcare services, Luque and colleagues (2008) concluded that distance, clinic hours of operation, and transportation costs were the primary obstacles. In addition to the financial burden, they also found that there was little understanding of symptoms related to respiratory infections, which delayed the pursuit of medical assistance and prolonged reliance on home remedies.

This work around Tungurahua inspired Whiteford and Tobin to develop their "cascade of effects" model of the consequences of disaster. The impacts (anticipated or otherwise) of disasters, displacement, and resettlement tend to "cascade" and perpetually undermine peoples' capacity to mitigate or effectively adapt to chronic hazards and post-disaster and resettlement conditions (Whiteford & Tobin, 2009). This framework was further developed and operationalized in a series of studies of social support networks in recovering communities and resettlements in Cantons Penipe (Chimborazo) and Cotaló (Tungurahua). American anthropologists A.J. Faas and Eric Jones led several of the team's investigations of social networks, social support, and wellbeing (e.g., Faas et al., 2014; Jones et al., 2013, 2014, 2015). One study examined gendered dynamics of formal (i.e., from institutions) and informal social support exchanges in three communities recovering from the 1999 and 2006 eruptions and two subsequent resettlements (Faas et al., 2014). The team found that, in most cases, men were more likely to be both givers and receivers of almost every type of support. Beyond identifying unequal access to scarce resources, Jones and colleagues (2014: p. 21) found that women were "expected to immediately adapt and perform all of their previous tasks in the same efficient manner, which means still bearing, raising, and caring for their children, as well as dealing with the effects of the resettlement". They found that anxiety, depression, and, in some cases, symptoms associated with posttraumatic stress disorder went untreated as women were forced to place priority on their household and familial obligations. By examining the relationship between social network structure and measures of wellbeing in disaster-affected populations in Ecuador (and, for comparison, Mexico), the study found that networks of varying kinds benefit genders differently. In cases where support was available to women post-disaster, factors such as resettlement, and variation in social network structure resulted in different support outcomes. However, in both cases social support and wellbeing outcomes consistently favored men over women.

Critical studies: minga *practice, expert practice*

A.J. Faas developed a study that focused on the role of traditional Andean cooperative labor parties, or *mingas*, in disaster recovery and resettlement in Cantón Penipe. He examined various aspects of *minga* practice in two resettlements for displaced *campesinos* and one displaced community where resettlers returned to recover livelihoods lost in resettlement. By focusing on the role of *minga* labor parties in these resettlements and villages where displaced *campesinos* returned to rebuild and cultivate, Faas was able to trace the production and contestation of what we might call "procedural vulnerability" (after Veland et al., 2013). The study called attention to the ways displaced *campesinos* organized to recover their livelihoods and sense of community, and the colonial legacies of *minga* practice as they played out in resettlements and recovering villages. Faas (2015) explored transitions of livelihoods and *minga* practice between resettlements in Penipe and Pusuca; the former a 287-house resettlement built by the Ecuadorian Ministry of Urban Development and Housing (MIDUVI) and US-based Christian disaster relief organization Samaritan's Purse on a landless urban grid, the latter a 42-house community constructed by Ecuadorian NGO Fundación Esquel with arable land for each household. The NGOs organized resettler *mingas* to construct resettlements (the state employed only contract labor), hoping to work through local cultural practices and promote a sense of ownership and pride in the resettlements. Several local leaders also learned that organizing *mingas* helped reflect community solidarity and attract the attention of NGOs in their home communities on the volcano. But the practice soon declined in Penipe, while it continued regularly in Pusuca and reactivated villages on the volcano, with projects including building repair, irrigation and potable water, and the clearing of roads and paths. Absent land or communal resources, there was no livelihood around which to organize *mingas* in Penipe and the power of local leaders – largely underwritten by their capacity to attract outside resources, which was itself buttressed by their capacity to organize *mingas* – diminished in the resettlement.

Faas (2017a) then examined *minga* practice as it varied between households in the Pusuca resettlement and a village, Manzano, where resettlers returned to make their livelihoods in the shadow of the volcano. He found that *minga* schedules being set by engineers employed by an NGO to design and manage work on an irrigation canal conformed to a 9–5 weekday schedule, often precluding the participation of those who had turned to wage labor to adapt to the scarcity of agricultural resources. Leaders in both communities sanctioned households for nonparticipation with fines of US$10–20 and loss of access to community resources (especially irrigation). But this power was exercised flexibly in Manzano, where villager schedules were accommodated, and inflexibly in Pusuca, where many villagers fell increasingly behind. Though some were excluded from *minga* practice in Manzano, this was largely due to internal conflicts and the redrawing of the relational boundaries of the community in resettlement.

Finally, Faas (2017b) studied the power dynamics of *minga* practice by situating *minga* as an enduring and contested feature of Andean statecraft at multiple levels of scale. Historically, *mingas* were both mutual aid practices and mechanisms of the state subordination and regulation of rural peripheries employed by the Incas, Spanish colonial administrators, and *hacendados* to extract free labor from indigenous and *campesino* groups. Following political reforms, social movements, and the proliferation of NGOs that characterized Ecuador in the late 20th and early 21st centuries, *minga* became a central feature of civil society, participatory governance, and development strategies, and part of the administrative ordering of society at the village level that enables local action and the formation of inter-community alliances where the state is notably absent (see Colloredo-Mansfeld, 2009: pp. 17–20). These dimensions of *minga* blended into each other in post-disaster resettlement as the practice was disembedded from local contexts and reembedded in the expert imaginations of humanitarian and resettlement agencies. Local economic strategies took on entrepreneurial characteristics (Faas, 2018) that articulate with neoliberal ambitions of state and global institutions; *campesino* ambitions and desires were produced and invoked as if they were locally derived, while at the same time being co-constructed by dominant interests. In sum, the cases considered in this study reveal the blurred distinctions between disciplines disaster-affected peoples impose on themselves and those to which they are subjected.

Ecuadorian development studies scholar Maria Teresa Armijos and colleagues (2017) and British geographer Roger Few and colleagues (2017) produced important studies of the livelihood impacts of the eruptions of Tungurahua, the ensuing resettlements, and adapting to chronic volcanic hazards. These studies are part of a global initiative, Strengthening Resilience in Volcanic Areas (STREVA), a four-year project funded by the Increasing Resilience to Natural Hazards Programme, a joint venture of the British Natural Environment Research Council and Economic and Social Research Council. In Ecuador, they collaborate with the Instituto Geofísico, whose scientists have engaged in monitoring and local-level risk reduction and management initiatives, including the recruitment of volunteers, known as *vigías* (lookouts). The *vigías* are roughly 35 citizen volunteers from areas all around Tungurahua who are trained, supported, and equipped with hand-held radios by the Instituto Geofísico since 2000 to be "the eyes and ears of the volcano" and report on volcanic activity to the Tungurahua Volcano Observatory and local officials. They report observations of volcanic activity, such as the number and strength of explosions and the volume, direction, color, and thickness of ashfall.

One of the most compelling and transformative products of the STREVA project in Ecuador has been a citizen science project with the *vigías*. This project entailed a series of reflective workshops, led by Armijos and Riobambeño artist Pablo Sanaguano, in which *vigías* were invited to share narratives, photos, and drawings of their experiences during, between, and since the major eruptions of Tunugrahua in 1999 and 2006. These narratives reflect the special and situated perspectives these citizen scientists bring to disaster risk reduction and management while calling attention to the increasingly locally salient idiom of *convivir* (to coexist) with the volcano (see also Few et al., 2017: p. 75). At the time of writing, the team is planning to publish a compilation of narratives, drawings, and photos in a book that will be made available in communities around Tungurahua, and considering a traveling exhibit to share their knowledge and experiences with others in Ecuador.

Critical geography, environmental studies, and risk management

French geographer Julien Rebotier's (2016) publication, *El riesgo y su gestión en Ecuador* (*Risk and Risk Management in Ecuador*), is the product of his extensive study of disaster prevention and management in Ecuador and is today the most comprehensive social science study of the topic in Ecuador. He is in one sense concerned with the production of environment and the factors that make a given society vulnerable, with attention to unequal power relations, the social construction of space, tensions between sustainable development and market production, risk as a rationale for intervention, and understanding all of these factors historically (Rebotier, 2016: pp. 33–38). Rebotier (2015) has called attention to unregulated construction on flatlands prone to flooding and the production of peripheral developments of vulnerable populations at a distance from health and employment resources. Rebotier is, above all, concerned with the production and occupation of space, human-environment relations, social organization, and politics as drivers of risk. He critically analyzes local case studies, the limitations of a hazard-centric orientation to risk and disaster, and the complex and extensive transformation of risk management in Ecuador.

One of the standout aspects of Rebotier's study is the discussion of the transformation of risk management agencies in Ecuador and the Andean region. Ecuador lacked specific legal frameworks and institutions for risk management through the 1950s, until the enactment of the Civil Defense Law in 1960 (Rebotier, 2016: p. 53). These operations remained principally focused on responses into the first years of the 21st century, with little in the way of substantive change until the 2006 eruptions of Tungurahua and the constitutional reforms of 2008. Important regional and multinational initiatives precipitated by the mega El Niño of 1997–1998 helped galvanize concerted efforts for disaster risk reduction in Ecuador at the turn of the century. In 2002, the governments of Bolivia, Colombia, Ecuador, Peru, and Venezuela formed the Proyecto de Apoyo a la Prevención de Desastres en la Comunidad Andina (The Support for Disaster Prevention in the Andean Community Project, PREDECAN) and, subsequently, the Comité Andino para la Prevención y Atención de Desastres (The Andean Committee for Disaster Prevention and Response, or CAPRADE) to institutionalize national risk management structures. Both entities became important resources for advancing policy and practice in disaster prevention in Ecuador and the broader Andean region (see Bermúdez et al., 2016). Rebotier is at times critical of these multinational efforts because they represent, after all, a "heterogeneous community, whose apparent unanimity is superficial . . . crossed by multiple tensions around risk challenges" (2016: p. 43).[7] In his analysis, because Ecuador as a state is part of this international community, adopting the transnational "movement around risk management" has constrained the implementation of risk management and public policies (2016: p. 43). Rebotier nonetheless notes the importance of connecting Ecuadorian risk management with regional and multinational efforts, as the topic of risk management has become "a vector of governance, an instrument of governance and a fundamental objective (consolidated by the debate on climate change) of international relations" (2016: p. 43).

Rebotier describes the process of the formation of the Secretaría Nacional de Gestion de Riesgos (SGR) that currently handles risk management in Ecuador as a result of two articles (389 and 390) in the revised Constitution of 2008. The SGR absorbed the functions of the Defensa Civil, which remains an operational arm of SGR, and the SGR mandate calls for developing and institutionalizing risk management in Ecuador in three key areas: a) creating normative standards based on national and global programs (e.g., the UN-brokered Sendai Framework for Disaster Risk Reduction), b) advancing the principles of *Buen Vivir*[8] to frame concrete actions for risk reduction at varying scales, and c) the coordination of operational priorities and capacities of relevant agencies, local committees, and technical-scientific institutes (Rebotier, 2016: pp. 48–49). Though it began as a Technical Secretariat, in November 2009, the category for SGR was raised to National Secretariat of Risk Management (SNGR). Since the adoption of the SNDGR in 2008, Rebotier suggests there have been advancements made in creating a more "risk-oriented approach to development and territorial planning to preparation and response" and since the creation of the SNGR in 2009, "its mandate has ceased to be disaster response" and has included "complex reflections on livelihoods and the quality of post-disaster recovery" (2016: p. 48). Despite these important developments, Rebotier points to several enduring challenges to more effective risk management in Ecuador. One key issue is fostering the development of local capacities, yet outside Ecuador's large urban centers, the vast majority of the 221 municipalities lack necessary financial and human resources, including technical staff and civil servants (Rebotier, 2016: p. 70).

Lorena Cajas, who led the SGR as it absorbed the Defensa Civil in 2007–2008, completed a thesis in socioenvironmental studies at FLACSO-Ecuador focusing on the incorporation of risk management into Ecuadorian development policy

(Cajas, 2010). Cajas (2012) considers the state as a generator of risk and vulnerability in Ecuador. She points to the evolution of the perception of disaster as a supernatural force, to a natural (i.e., hazard-centric) force, to the concept of vulnerability as it is commonly engaged in the critical social science of disasters in recent decades. Cajas argues that household level and community level risk/ vulnerability reduction alone are inadequate unless tied to more comprehensive, intersectoral plans. After considering the failed attempts to rebuild the Baños-Riobamba road alongside the volcano Tungurahua, Cajas makes the point that the state has the responsibility to coordinate, integrate, and provide resources to otherwise disparate efforts and ensure their sustainability. Such resources include technical assistance, training, and contingency planning in order to create a "culture of prevention". The development of the SGR was a positive step, as the Civil Defense had been underfunded, focused exclusively on response, and not given any decision-making power in terms of disaster risk reduction and management. Cajas places emphasis on public information and the promotion of techno-scientific institutions in risk reduction and management. She is concerned that, until the state absorbs the roles of risk reduction and management into core mandates of all agencies and central elements of development strategies, it will remain a generator of risk and vulnerability.

Cajas and (especially) Rebotier engage with the critical approach to risk management and disasters in Latin America associated with the work of García-Acosta (2018), Lavell (1993), and Maskrey (1993), which entails a focus on development, poverty, injustice, and politics when working on the social production of disasters and disaster management. That is, they draw heavily upon and contribute to many of the core conversations in the Anthropology of Disasters and serve as excellent examples of the porosity of disciplinary boundaries – especially at the interstices of anthropology, geography, and environmental studies, and most especially in the social sciences of disaster. Moreover, Cajas and Rebotier both carefully study the policy spheres and institutional dynamics of disaster risk reduction, response, and recovery that, while clearly significant in the broader field of the Anthropology of Disasters (e.g., Revet & Langumier, 2015), are somewhat underrepresented in much of the research on disasters in Ecuador.

Horizons: extraction and climate change

There are important hazards issues on the horizon in Ecuador, and deforestation and climate change are chief among them. Anthropological research on deforestation in Ecuador began in the 1970s. Ecuadorian anthropologist Eulalia Carrasco (1983) studied among the Chachi people of the rainforest region of coastal Esmeraldas with attention to resistance to timber extraction that caused ecological devastation and the spread of river blindness (*onchocerciasis*) that decimated the Chachi population. American environmental sociologist Diane C. Bates (2008) has written extensively about deforestation in Ecuador, calling attention to the fact that, while Amazonian deforestation has received the most attention in global environmental debates, "the Pacific and Andean forests in Ecuador are currently smaller and arguably more endangered than the Amazonian forests" (Bates, 2008: p. 260). Bates's work also highlights the cross-cutting production of hazard vulnerability in pointing to housing shortages and poverty in Guayaquil leading the poor to construct houses on wetlands, which contributes to mangrove deforestation and increased risk of flooding in these poor communities (Bates, 2008: pp. 259–260).

Climate change impacts such as melting glaciers and snow are moving apace and expected to continue to affect Ecuadorian communities and environments in the coming years. Cotopaxi has lost more than a third of its glacier since 1976, and other lower mountains could lose all glaciers in the coming decades. Reduced snowpack could pose a serious threat to highland communities and cities that rely on this resource for their water supply (Schoolmeester et al., 2018). The poignant term "CO_2lonialism," was coined by the Indigenous Environmental Network in 2007 and first discussed in scholarship by American geographer Julianne Hazelwood (2010). The term is employed to theoretically synthesize the intimate relationships between capitalism, colonialism, and climate change. Hazelwood's work focuses on palm oil cultivation and agrofuel production in San Lorenzo in the coastal province of Esmeraldas and the environmental impacts – deforestation, poisoned rivers – and social conflicts that resulted from the growth of these industries. She encourages readers to think beyond development in narrowly economic terms to focus on livelihoods, climate, and environmental sustainability. Importantly, she focuses on Afro-Ecuadorian, Chachi, and Awá communities directly affected by extractive economies and how they create and maintain "geographies of hope" and aspire to the good and self-determination, invoking the Quichua idiom of *sumak kawsay* ("*buen vivir*" or "good living") that has become a structuring metaphor of pan-indigenous and *campesino* social movements in the Andes.

Discussion and conclusions: the state of Anthropology of Disasters within Ecuador

The state of the art of the Anthropology of Disasters in Ecuador is today decidedly promising and, in the course of researching and writing this chapter, I have come to recognize it as less fragmented than I had earlier thought; still, many apertures remain. Put succinctly, disasters and a wide range of environmental hazards feature prominently in Ecuadorian histories, presents, and futures. Yet, many lacunae remain in anthropological and other critical social science treatments of these processes. Though the topic of disasters as a focused research agenda is virtually untouched by Ecuadorian anthropologists, there is increasing attention to the topic in publications and theses produced by students, faculty, and professionals associated with FLACSO-Ecuador.[9] It is worth mentioning that much of this new thesis work has taken place under the direction of Teodoro Bustamante, research professor at FLACSO-Ecuador, whose work addresses issues of resource management and oil exploitation. In a line of thinking that might contribute to anthropological thinking on disaster, Bustamante (2010) has called for a reconsideration of how we think of nature by reflecting critically on Renaissance thought on the topic. He suggests a reckoning with both reason and the supernatural brought on by accelerated environmental, social, and political crises of the 21st century.

Several publications in the relatively new (as of 2008) journal, *Letras Verdes*, run by master's students in the FLACSO socioenvironmental studies program, have addressed the topics of risk, hazards, and disasters (see Estacio & Narváez, 2012; Vallejo & Vélezo, 2009). And a 2010 special issue of *Íconos*, the multidisciplinary social science journal published by FLACSO-Ecuador since 1997, was edited by Franklin Ramírez y Hugo Jácome and devoted to "nature and capitalist crisis". Though it featured no anthropologists, multiple social scientists considered the relationships between environmental degradation and capitalist processes of extraction, production, and consumption. At the time of writing, *Íconos* is in the process of producing a new special issue for early 2020, "Community, vulnerability, and reproduction in disaster conditions", edited by Cristina Vega, Ana Gabriela Fernández, and Johannes Waldmüller.

Like much of the ethnography in Ecuador, the Anthropology of Disasters shares a concern with history and memory and these shared foci offer special opportunities for work synthesizing these areas of research. For instance, several cities in Tungurahua province were destroyed by massive earthquakes in 1949, and then again in 1987 (Latrubesse, 2010), yet to date there are no focused anthropological studies on the causes and impacts of these disasters. However, American anthropologist Rachel Corr (2010) conducted an extensive study of contemporary ritual practices in the highland indigenous community of Salasaca, with attention to the ways in which these practices were shaped by large-scale historical practices and cultural encounters and how they were transformed at the local level among individual practitioners. Though Corr is not concerned with theorizing hazards and disasters, she captures the ways symbolism and ritual practice are intimately tied to the Salasacan landscape and reveals how local memories are often entangled with past eruptions and disasters. According to one local narrative, God caused an earthquake from which emerged mountains and rocks that produced the Catholic saints, which is said to account for the enduring contemporary importance of mountain worship for access to sacred powers (Corr, 2010: p. 55). The 1949 earthquake not only toppled buildings in Chimborazo and Tungurahua Provinces and resulted in the deaths of just over 5,000 people, it also triggered multiple landslides that extended disaster impacts to approximately 100,000 people (USGS, n.d.). It should therefore come as little surprise that the event remains salient in Salasacan memories and the ways locals mark historical time. One of Corr's (2010: p. 1) local indigenous collaborators recalled the trauma of the event and the importance of gathering at a local sacred mountain in the aftermath to pray. Locals and the former bishop of Ambato also recalled that three nuns came to Salasaca from Imbabura Province to the south in the days following the quake after several nuns, priests, and seminarians died in the disaster (Corr, 2010: pp. 1, 50-51). The church and convent that were later constructed became important features of the local built environment that likewise conjure memories of the 1949 disaster. Drought too occupies a special place in local memories and had, in the past, inspired recourse to the sacred mountain, Quinchi Urcu (Corr, 2010: p. 142). One man recounted a tale from his grandfather, who had told him of going to Quinchi Urcu with flutes, drums, hominy, and maize beer (*chicha*) to implore the mountain to bring rain. The storyteller recalled that this had been a *minga* of sorts in which all locals were called upon to bring food to make a festive rite at the mountain (Corr, 2010: p. 150).

These cases point to important opportunities to investigate further the ways in which disasters of the past endure in social memory and influence meanings, material arrangements, practices, and discourses in the present. The opportunities to begin synthesizing these areas of research are too many to name. Just as citizens began recovering from devastating landslides that killed more than 50 residents near Quito in 1966, a major drought claimed many lives in 1968. How do we understand how these processes continued to shape urban lives and environments since then? What practices in the present are rooted in these past experiences?

I should also point to opportunities to study, not merely "local" responses to hazards and disasters, but to the intersections of powers, practices, and knowledges that are surfaced in disaster dramas. The work of both Sawyer and Wasserstrom points to how disasters are not just opportunities for dominant power but also structuring idioms and catalysts for subaltern resistance. Faas points to the contested terrains of *minga* practice as local *campesinos* use *minga* for their own self-driven recovery strategies and to court outside resources and relationships from institutions who likewise make their marks on *minga* practice by using it to impose their own objectives and disciplines. Armijos, Few, and colleagues have likewise opened an important subject for the study of varieties of disaster science and technology, such as an investigation of the intersection of knowledges between the Instituto Geofísico and local *vigías* who live around the volcano. Rebotier, Faas, Sawyer, and Hazelwood all identify opportunities to better situate local responses in national, regional, and global processes. And, within Ecuador as in the broader Andean region, there have long been intimate connections between ethnographic, archaeological, and ethnohistorical work, which speaks to the enduring importance of work like the studies by Taylor and Manzanilla and their capacity to teach us much about the present and future, as well as the past.

I resist offering a programmatic agenda beyond encouraging further drawing together the rich ethnographic work in and of Ecuador and the critical and applied Anthropology of Disasters. Today, there are some significant studies that are not in dialog with global conversations about the Anthropology of Disasters. Likewise, there are studies of risk, hazards, and disasters in Ecuador that engage with global conversations, but do so insufficiently with the rich body of Ecuadorian anthropology. There is plenty of room for growth in both directions to put these vast bodies of literature in conversation in future studies.

Notes

1 I would like to thank Daniel Maldonado for his contributions as a research assistant. The San José State University Emeritus and Retired Faculty Association provided generous funding for research related to this chapter. I am forever grateful to Linda Whiteford,

Graham Tobin, Art Murphy, Eric Jones, and Hugo Yepes for guiding and supporting my entry into anthropology and disasters in Ecuador. And I especially appreciate Virginia García-Acosta for her insight and guidance on this project. Y muchisimas gracias a mis amigos de Penipe por su amistad y su valiosa colaboración, dios les pague.

2 It has become standard vocabulary in disaster studies to isolate *hazards* and *risk* from *disaster*, and to distinguish *natural* (e.g., hydrometeorological, geophysical) from *technological* (e.g., nuclear, toxic release) hazards. I prefer the term *environmental* hazards for what otherwise might be called *natural*, as the distinction between "natural" and anthropogenic has become increasingly untenable (e.g., Latour, 2004).

3 I have attempted to recognize and provincialize the influence of North American and European scholarly engagements with the problems of risk, hazards, and disasters in Ecuador by situating them in world anthropologies and naming the nationalities of the scholars referenced. My objective was to foreground to the extent possible Ecuadorian and Latin American scholars working in areas related to the theme of this chapter. Though this is a rather blunt approach, my intent is not to reify nationality at the expense of the intersectionality of identity. I recognize that many scholars were born or educated in one national context, yet have worked for years (in some cases, decades) in another. I do not mean to undermine this by pigeonholing them in one national identity.

   Disciplinary boundaries too are difficult to sustain – especially in the cases of Armijos, Few, García-Acosta, and several others. But, given the concern of this volume with anthropology, I have preserved disciplinary labels while including plenty of work by scholars in related disciplines that should not be bracketed out of consideration in such a review as this. I recognize the limitations and tradeoffs of these compromises and have endeavored to avoid reifying superficial identities.

4 In his history of Ecuadorian anthropology, Segundo Moreno Yánez (1992: pp. 43–44) makes only passing mention of Haro, while noting that the authors on whose work Haro based much of his historical compilations – Aquiles Pérez Tamayo and Jacinto Jijon y Caamaño – drew conclusions from flawed linguistic and archaeological data. While Haro's work reflects incredible breadth and depth in reading, he relied heavily on these sources and the works of Spanish chroniclers. He also tended to write rather condescendingly of indigenous religion and society as inferior to his own (see Haro, 1977: p. 91).

5 For an overview, see Bermúdez and colleagues (2016); see also key works such as Santillán (2015) and Kingman (2006).

6 Established at the Escuela Politécnica Nacional in Quito in the 1980s to monitor seismic and volcanic processes.

7 All translations are mine, unless otherwise noted.

8 *Buen vivir* (Quechua *sumac kawsay*) is a principal of wellbeing over profit that emerged from Bolivian politics to become metaphoric principle of trans-Andean social movements and policies that emphasize solidarity, reciprocity, gender equity, and environmental sustainability.

9 See Cruz Jiménez (2011) for socioenvironmental studies and López Paredes (2015) for anthropology.

References

Armijos, M. T., Phillips, J., Wilkinson, E., Barclay, J., Hicks, A., Palacios, P., Mothes, P. & Stone, J. (2017) Adapting to Changes in Volcanic Behaviour: Formal and Informal Interactions for Disaster Risk Management at Tungurahua Volcano, Ecuador. *Global Environmental Change*. 45, pp. 217–226.

Baez Ullberg, S. (2017) La Contribución de la Antropología al Estudio de Crisis y Desastres en América Latina. *Iberoamericana-Nordic Journal of Latin American and Caribbean Studies*. 46 (1), pp. 1–5.

Bankoff, G. (2001) Rendering the World Unsafe: "Vulnerability" as Western Discourse. *Disasters*. 25 (1), pp. 19–35.

Bates, D. C. (2008) Deforestation in Ecuador. In: Torre, C. de la & Striffler, S. (eds.). *The Ecuador Reader: History, Culture, Politics*. Durham, Duke University Press, pp. 257–266.

Bermúdez, N., Cabrera, S., Carrión, A., Hierro, S. del, Echeverría, J., Godard, H. & Moscoso, R. (2016) La investigación urbana en Ecuador (1990–2015) cambios y continuidades. In: Metzger, P., Rebotier, J., Robert, J., Urquieta, P. & Vega Centeno, P. (eds.). *La Cuestión Urbana en la Región Andina: Miradas sobre la investigación y la formación*. Quito, Pontificia Universidad Católica del Ecuador Centro de Publicaciones, pp. 117–174.

Blaikie, P., Cannon, T., Davis, I. & Wisner, B. (1994) *At Risk: Natural Hazards, People's Vulnerability and Disasters*. New York, Routledge.

Boelens, R. (2015) *Water, Power, and Identity: The Cultural Politics of Water in the Andes*. New York, Routledge.

Bustamante, T. (2010) El Pensamiento Sobre la Naturaleza en el Renacimiento. *Antropología: Cuadernos de Investigación*. 10, pp. 61–73.

Bustamante, T. & Jarrín, M. C. (2005) Impactos Sociales de las actividades petroleras en el Ecuador: Un análisis de los indicadores. *Íconos*. 21, pp. 19–34.

Cajas, L. (2012) El Estado como generador de riesgos: el caso de Ecuador. *Letras Verdes*. 11, pp. 64–72.

Cajas, L. (2010) *La incorporación de la gestión de riesgos como una política de desarrollo en el quehacer institucional público, el caso del Ecuador*. Masters thesis. Quito, FLACSO.

Carrasco, E. (1983) *El Pueblo Chachi: El Jeengume Avance*. Quito, Ediciones Abya-Yala, Colección Ethnos.

Clark, A. K. (2002) The Language of Contention in Liberal Ecuador. In: Lem, W. & Leach, B. (eds.). *Culture, Economy, Power: Anthropology as Critique, Anthropology as Praxis*. New York, State University of New York Press, pp. 150–162.

Clark, A. K. (1998) *The Redemptive Work: Railway and Nation in Ecuador, 1895–1930*. Lanham, Rowman & Littlefield.

Colloredo-Mansfeld, R. (2009) *Fighting Like a Community: Andean Civil Society in an Era of Indian Uprisings*. Chicago, Chicago University Press.

Corr, R. (2010) *Ritual and Remembrance in the Ecuadorian Andes*. Tucson, University of Arizona Press.

Cruz Jiménez, M. A. (2011) *Vulnerabilidad Social de los Escenarios Volcánicos en los Andes Centrales del Ecuador. Análisis de la Gestión del Riesgo Caso Puela – Provincia De Chimborazo*. Masters thesis. Quito, FLACSO.

Estacio, J. & Narváez, N. (2012) Incendios forestales en el Distrito Metropolitano de Quito (DMQ): Conocimiento e intervención pública del riesgo. *Letras Verdes*. 11, pp. 27–52.

Faas, A. J. (2018) Petit Capitalisms in Disaster, or the Limits of Neoliberal Imagination: Displacement, Recovery, and Opportunism in Highland Ecuador. *Economic Anthropology*. 5 (1), pp. 32–44.

Faas, A. J. (2017a) Enduring Cooperation: Enduring Cooperation: Time, Discipline, and Minga Practice in Disaster-Induced Displacement and Resettlement in the Ecuadorian Andes. *Human Organization*. 76 (2), pp. 99–108.

Faas, A. J. (2017b) Reciprocity and Vernacular Statecraft: Changing Practices of Andean Cooperation in Post-Disaster Highland Ecuador. *The Journal of Latin American and Caribbean Anthropology*. 22 (3), pp. 495–513.

Faas, A. J. (2016) Disaster Vulnerability in Anthropological Perspective. *Annals of Anthropological Practice* 40 (1), pp. 9–22.

Faas, A. J. (2015) Disaster Resettlement Organizations, NGOs, and the Culture of Cooperative Labor in the Ecuadorian Andes. In: Companion, M. (ed.). *Disaster's Impact on*

Livelihoods and Cultural Survival: Losses, Opportunities, and Mitigation. Boca Raton, CRC Press, pp. 51–62.

Faas, A. J. & Barrios, R. (2015) Applied Anthropology of Risk, Hazards, and Disasters. *Human Organization.* 74 (4), pp. 287–295.

Faas, A. J., Jones, E., Whiteford, L., Tobin, G. & Murphy, A. (2014) Gendered Access to Formal and Informal Resources in Postdisaster Development in the Ecuadorian Andes. *Mountain Research and Development.* 34 (3), pp. 223–234.

Few, R., Armijos, M. T. & Barclay, J. (2017) Living with Volcán Tungurahua: The Dynamics of Vulnerability During Prolonged Volcanic Activity. *Geoforum.* 80, pp. 72–81.

García-Acosta, V. (2018b) Vulnerabilidad y desastres. Génesis y alcances de una visión alternativa. In: González de la Rocha, M. & Saraví, G. A. (coord.). *Pobreza y Vulnerabilidad: debates y estudios contemporáneos en México.* Mexico, CIESAS, pp. 212–239.

García-Acosta, V. (2005) El riesgo como construcción social y la construcción social de riesgos. *Desacatos. Revista de Antropología Social.* 19, pp. 11–24.

García-Acosta, V. (1993) Enfoques teóricos para el estudio histórico de los desastres naturales. In: Maskrey, A. (comp.). *Los desastres no son naturales.* Bogota, LA RED, Tercer Mundo Editores, pp. 155–166.

González Suárez, F. (1890) *Historia General de la República del Ecuador,* vol. 1. Biblioteca Virtual Universal.

Haraway, D. J. (2016) [2003] The Companion Species Manifesto: Dogs, People, and Significant Otherness. In: Wolfe, C. (ed.). *Manifestly Haraway.* Minneapolis, University of Minnesota Press, pp. 91–198.

Haro Alvear, S. L. (1977) *Puruhá, Nación Guerrera.* Quito, Editora Nacional.

Haro Alvear, S. L. (1973) Montañas Sagradas del Reino de Quito. Quito: *Boletín de la Academia Nacional de Historia,* pp. 121–130.

Hazelwood, J. A. (2010) Más allá de la Crisis Económica: CO_2lonialismo y Geografías de Esperanza. *Íconos.* 36, pp. 81–95.

Jones, E. C., Faas, A. J., Murphy, A. D., Tobin, G. A. & Whiteford, L. M. (2013) Cross-Cultural and Site-Based Influences on Demographic, Individual Wellbeing, and Social Network Factors Predict Risk Perception in Hazard and Disaster Settings in Ecuador and Mexico. *Human Nature.* 24 (1), pp. 5–32.

Jones, E. C., Murphy, A. D., Faas, A. J., Tobin, G. A., McCarty, C. & Whiteford, L. M. (2015) Post-Disaster Reciprocity and the Development of Inequality in Personal Networks. *Economic Anthropology.* 2 (2), pp. 385–404.

Jones, E. C., Tobin, G., McCarty, C., Whiteford, L., Murphy, A., Faas, A. J. & Yepes, H. (2014) Articulation of Personal Network Structure with Gendered Well-Being in Disaster and Relocation Settings. In: Roeder Jr., L. W. (ed.). *Issues of Gender and Sexual Orientation in Humanitarian Emergencies, Humanitarian Solutions in the 21st Century.* Switzerland, Springer International Publishing, pp. 19–31.

Kimerling, J. (1993) *Crudo Amazónico.* Quito, Abya-Yala.

Kingman, E. (2006) *La ciudad y los otros: Quito 1860–1940.* Quito, FLACSO.

Krupa, C. & Nugent, D. (2015) Off-Centered States: Rethinking State Theory Through an Andean Lens. In: Krupa, C. & Nugent, D. (eds.). *State Theory and Andean Politics: New Approaches to the Study of Rule.* Pennsylvania, University of Pennsylvania Press, pp. 1–32.

Kulstad-González, T. & Faas, A. J. (2016) Afterword: Preparing for Uncertainties. *Annals of Anthropological Practice.* 40 (1), pp. 98–105.

Lane, L. R., Tobin, G. A. & Whiteford, L. M. (2003) Volcanic Hazard or Economic Destitution: Hard Choices in Baños, Ecuador. *Environmental Hazards.* 5, pp. 23–34.

Larson, B. (2008) Andean Highland Peasants and the Trials of Nation Making During the Nineteenth Century. In: Salomon, F. & Schwartz, S. B. (eds.). *The Cambridge History of the Native Peoples of the Americas, Vol. 3.2: South America.* New York, Cambridge University Press, pp. 558–703.

Latour, B. (2004) *Politics of Nature: How to Bring the Sciences into Democracy.* Cambridge, Harvard University Press.

Latrubesse, E. (2010) *Natural Hazards and Human-Exacerbated Disasters in Latin America.* Amsterdam, Elsevier.

Lavell Thomas, A. (1993) Ciencias sociales y desastres naturales en América Latina: un encuentro inconcluso. In: Maskrey, A. (comp.). *Los desastres naturales no son naturales.* Bogotá, LA RED, Tercer Mundo Editores, pp. 135–154.

López Paredes, G. G. (2015) *Memoria del Desastre: Afectados por las Inundaciones Provocadas por los Desbordamientos del Río Bulubulu en el Cantón El Triunfo, Caserío Payo Chico.* Master's thesis. Quito, FLACSO.

Luque, J. S. (2007) Healthcare Choices and Acute Respiratory Infection: A Rural Ecuadorian Case Study. *Human Organization.* 66 (3), pp. 282–291.

Luque, J. S., Whiteford, L. M. & Tobin, G. A. (2008) Maternal Recognition and Healthcare-Seeking Behavior for Acute Respiratory infection in Children in a Rural Ecuadorian County. *Maternal and Child Health Journal,* 12, pp. 287–297.

Macías, J. M. & Aguirre, B. E. (2006) A Critical Evaluation of the United Nations Volcanic Emergency Management System: Evidence from Latin America. *Journal of International Affairs.* 59 (2), pp. 43–61.

Manzanilla, L. (1997) Indicadores Arqueológicos de Desastres: Mesoamérica, Los Andes, y Otro Casos. In: García-Acosta, V. (coord.). *Historia y Desastres en América Latina.* Lima, LA RED, CIESAS, Tercer Mundo Editores, vol. II, pp. 33–54.

Maskrey, A. (comp.). (1993) *Los desastres no son naturales.* Bogotá, LA RED, Tercer Mundo Editores.

Moreno Yánez, S. (1992) *Antropología Ecuatoriana: Pasado y Presente.* Quito, EDIGUIAS C. LTDA.

Moseley, M. E. (1999) Convergent Catastrophe: Past Patterns and Future Implications of Collateral Natural Disasters in the Andes. In: Oliver-Smith, A. & Hoffman, S. M. (eds.). *The Angry Earth: Disaster in Anthropological Perspective.* New York, Routledge, pp. 55–71.

Muratorio, B. (1998) *Rucuyaya Alonso y La Historia Social y Económica del Alto Napo,* 2nd edition. Quito, Abya-Yala.

Oliver-Smith, A. & Hoffman, S. M. (eds.). (1999) *The Angry Earth: Disaster in Anthropological Perspective.* London, Routledge.

Prieto, M. (2015) *Estado y colonialidad. Mujeres y familias Kichwas de la Sierra del Ecuador, 1925–1975.* Quito, FLACSO.

Rachowiecki, R. (2008) Mountaineering on the Equator: A Historical Perspective. In: Torre, C. de la & Striffler, S. (eds.). *The Ecuador Reader: History, Culture, Politics.* Durham, Duke University Press, pp. 148–154.

Rebotier, J. (2016) *El riesgo y su gestión en Ecuador: Una mirada de Geografía social y política.* Quito, PUCE, Serie de Publicaciones en Ciencias Geográficas.

Rebotier, J. (2015) Dimensiones Sociales de los Riesgos y su Generación. In: Estrella, M. E. (ed.). *Serie Reflexiones Académicas, La Vulnerabilidad y los Riesgos, Estudios de Casos en el Ecuador.* Quito, CMYK Imprenta, pp. 13–23.

Revet, S. & Langumier, J. (eds.). (2015) *Governing Disasters: Beyond Risk Culture.* New York, Palgrave Macmillan.

Salgado Peñaherrera, G. (1999) Las Claves Para El Futuro. *Íconos*, pp. 68–77.

Santillán Cornejo, A. (2015) Imaginarios urbanos y segregación socioespacial: Un estudio de caso sobre Quito. *Cuadernos de Vivienda y Urbanismo*. 8 (16), pp. 246–263.

Sawyer, S. (2008) Suing ChevronTexaco. In: Torre, C. de la & Striffler, S. (eds.). *The Ecuador Reader: History, Culture, Politics*. Durham, Duke University Press, pp. 321–328.

Sawyer, S. (2004) *Crude Chronicles: Indigenous Politics, Multinational Oil, and Neoliberalism in Ecuador*. Chapel Hill, Duke University Press.

Schoolmeester, T., Johansen, K. S., Alfthan, B., Baker, E., Hesping, M. & Verbist, K. (2018) *Atlas de Glaciares y Aguas Andinos. El impacto del retroceso de los glaciares sobre los recursos hídricos*. Arendal, Norway, UNESCO & GRID-Arendal.

Sun, L. & Faas, A. J. (2018) Social Production of Disasters and Disaster Social Constructs: An Exercise in Disambiguation and Reframing. *Disaster Prevention and Management*. 27 (5), pp. 623–635.

Taylor, S. (2015) The Construction of Vulnerability along the Zarumilla River Valley in Prehistory. *Human Organization*. 74 (4), pp. 297–307.

Tobin, G. A. & Whiteford, L. M. (2002) Community Resilience and Volcano Hazard: The Eruption of Tungurahua and Evacuation of the Faldas in Ecuador. *Disasters*. 26 (1), pp. 28–48.

Tsing, A. L. (2005) *Friction: An Ethnography of Global Connection*. Princeton, Princeton University Press.

United States Geological Survey (USGS). (n.d.) *Today in Earthquake History*, August 5. Accessed May 3rd 2017.

Vallejo, A. & Vélezo, J. A. (2009) La percepción del riesgo en los procesos de urbanización del territorio. *Letras Verdes*, 3, pp. 29–37.

Veland, S., Howitt, R., Dominey-Howes, D., Thomalla, F. & Houston, D. (2013) Procedural Vulnerability: Understanding Environmental Change in a Remote Indigenous Community. *Global Environmental Change*. 23 (1), pp. 314–326.

Wasserstrom, R. (2016) Waorani Warfare on the Ecuadorian Frontier, 1885–2013. *The Journal of Latin American and Caribbean Anthropology*. 21 (3), pp. 497–516.

Whiteford, L. M. & Tobin, G. A. (2009) If the Pyroclastic Flow Doesn't Kill You, the Recovery Will: Cascading Impacts of Mt. Tungurahua's Eruptions in Rural Ecuador. In: Jones, E. C. & Murphy, A. D. (eds.). *The Political Economy of Hazards and Disasters*. Lanham, Altamira, pp. 155–178.

Whiteford, L. M. & Tobin, G. A. (2004) Saving Lives, Destroying Livelihoods: Emergency Evacuation and Resettlement Policies. In: Castro, A. & Springer, M. (eds.). *Unhealthy Health Policies: A Critical Anthropological Examination*. Walnut Creek, Altamira, pp. 189–202.

6 The Mexican vein in the Anthropology of Disasters and Risk

Virginia García-Acosta

Introduction: Mexican anthropology and the birth of Mexican Anthropology of Disasters[1]

Globally, anthropologists started to study disasters by the mid-20th century. These were studies developed mainly within the bounds of British anthropology. However, since the 1970s, the participation of anthropologists coming mainly from the United States is noticeable.

Mexican anthropology emerged and developed from the combination of three constituent factors: a national project, dominant paradigms, and their influence, and an institutional framework whose construction was strengthened by multiple national and international supports (Bueno, 2010). It did so at the beginnings of the 20th century; during the following five decades, it was dedicated almost exclusively to study past and contemporary Indigenous populations in a close relationship with history and archaeology. During the second half of that century, its interests began including other social groups and dabbled in a growing range of subject areas (Krotz, 2018).

Even though its origins can be traced to the beginnings of the 20th century, it ventured into this academic field in the last decades of that century. Some empirical anthropological and ethnological works carried out during the second half of the 20th century at Chiapas and Yucatán, although not motivated by a specific interest in the study of disasters, made such valuable contributions that we can recognize them as the seed of the field we now call "Anthropology of Risk and Disasters in Mexico". As will be seen later, 1985 represents a landmark in this new field of research. The presence of a disaster associated with a natural hazard, in this case the big earthquake in Mexico City, played a crucial role. From then on, systematic ethnographic research was developed, particularly at CIESAS (Centre for Research and Advanced Studies in Social Anthropology). This is also a major moment in the development of theoretical and methodological research on risks and disasters in Mexico, a field that has continued and expanded significantly.

Historical research related to this specialized field also began at that time, concomitant to an intense search for theoretical tools that would address the problem from a social perspective. The diachronic study and comparison of droughts associated with agricultural crisis and epidemics which occurred during the colonial era was an important precedent, as was the interest of archaeologists, ethnohistorians, and anthropologists in the study of floods. The latter have been present throughout Mexican history, as we will see later.

Advances in the anthropological study of disasters in Mexico were very important in the 1990s. A crucial step was the foundation in 1992 in Costa Rica, within the framework of the UN International Decade for Natural Disaster Reduction, of the Red de Estudios Sociales en Prevención de Desastres en América Latina (Social Research Network for Disaster Prevention in Latin America), better known as LA RED. It was a fruitful initiative that emerged from discussions among Latin American researchers, including several Mexican social scientists. It allowed researchers to carry out comparative and empirical research with other Latin American and Latin Americanists scholars, including anthropologists, who were unsatisfied with the analytical frameworks developed in the global north. In the Mexican case, disasters linked to hurricanes, such as Pauline (1997) and Isidore (2002), fueled Mexican anthropological risk and disaster research. In fact, the progress Mexico achieved throughout the last decades in the Anthropology of Disasters led it to be recognized as a sort of "spearhead" in the Latin American region.

The school of cultural ecology that derives from the multilinear evolution approach (Steward, 1955) first influenced Mexican disaster studies; it also followed a methodology focused on the combination of synchronic and diachronic dimensions, comparison, and a *longue durée* perspective (Braudel, 1958).

Anthropological and historical research carried out in Mexico and other Latin American regions has sustained, for example, the recognition that what was called "natural disasters" are not the result of the sole presence of natural phenomena, but the product of correlations between natural hazards and conditions of vulnerability, which are compounded by a growing social construction of risk. Social research in the global south has made evident that most disasters are not the result of external but internal processes: life conditions due to social and economic vulnerability are the main causes of disasters. Contributions from anthropology to the study of these issues have allowed researchers to identify variables previously not very visible, and even ignored, in risk management. The most important works have been carried out hand in hand with historical research. It became clear that disasters are processes that demonstrate historical analysis is essential in disaster research.

In this chapter, I will present an overview of the development, progress, and current state-of-the-art tools of the anthropological study of risk and disasters in Mexico, almost four decades after the publication of the pioneer works aforementioned and of the contribution the 1985 earthquakes triggered. The latter was a turning point in the development of an anthropological field of research, now called the sociohistorical study of risk and disasters. I will also consider the achievements made in publications and in undergraduate and graduate theses. National and international academic meetings will be addressed as well. I will make special emphasis on the contributions made by social anthropology to the study and understanding of risk and disasters in the theoretical and methodological level. These include the proposal of new approaches and concepts resulting

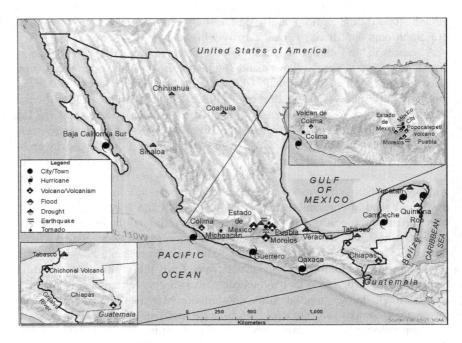

Map 6.1 Map Mexico. Case studies and main areas mentioned

from specific studies in which fieldwork is a constant variable, often in combination with archival and documentary research.

A disaster as a trigger of a new academic field

Several empirical anthropological publications are the basis of what we may call now Anthropology of Risks and Disasters in Mexico. The first one appeared after the 1982 Chichonal volcano eruption in Chiapas, the most devastating one during the last 100 years until the Pinatubo eruption in 1991. That event set off the interest of a group of Mexican anthropologists, whose resulting publication is currently recognized as the first contribution to the field (Báez-Jorge et al., 1985). They had as their main objective to evaluate, with a critical sense, the social effects generated by the disaster within the Zoque indigenous population living around the volcano. With a definite emphasis on the context in which the eruption occurred, thanks to the existence of previous historical and ethnographic studies in the region, the conclusions to which the authors arrived relate to what later we called "vulnerability and risk conditions". The unequal distribution of resources, underemployment, migration, and intra-ethnic conflicts all constituted the active elements of the disaster. Moreover, the locals knew the ancient volcano behavior since this was an indissoluble component of the zoque culture, and they had provided warnings that prognosticated its next eruption, but they were neglected by the authorities. The disaster happened.

The second example comes from late Canadian anthropologist Herman Konrad's analysis of the role of tropical storms between pre-Hispanic and contemporary Maya, also published in 1985. He highlights what he calls "the hurricane factor [that was] significantly overlooked" among environmental factors in pre-Hispanic Maya developmental processes. He considered these as endogenous elements of Maya culture. In this paper, as well as in another one published 11 years later (Konrad, 1996), he included data coming from pre-Hispanic and contemporary Maya on the Yucatan Peninsula examining hurricanes as trigger mechanisms in Maya history related to settlement patterns, subsistence patterns, migrations and demographic stability, warfare, and trade (1985). Konrad even identified what he called "Maya strategies of coping with environmental restraints and opportunities" which allow the minimization of not only the damage caused by recurrent tropical cyclones, but also for communities to profit from them.

The research projects that characterize our work tend to be attached to plans and projects designed in advance; nevertheless, on certain occasions reality presents problems that emerge suddenly, requiring our immediate attention. This happened in Mexico after the eruption of the Chichonal volcano in 1982, and especially in CIESAS after the disaster associated with the September 1985 earthquake.

Only 12 years after the founding of CIESAS,[2] its Director archaeologist Eduardo Matos and its Technical Consulting Council called upon the institution's researchers to temporarily suspend their medium and long-term research in order to attend, from the perspective of their areas of specialization, the study of the problems that resulted from the disaster. The response was overwhelmingly positive. Nearly 40 researchers, students, and collaborators participated in the initiative, carrying out focused short-term projects that were initiated in September of 1985 and which, in many cases, did not extend beyond six months. Generally, such studies attempted to articulate two basic issues: "an interest in making a critical assessment of the repercussions of the earthquakes on key sectors and the need to make an immediate assessment of the problems at hand" (García-Acosta, 2018e).

During the year and a half that followed, four books were published as part of the CIESAS series "Cuadernos de la Casa Chata" ("Notebooks of the Casa Chata")[3] which were pioneers of the social science perspective on disasters and what is today called the "global south". One of these books was the precursor of a new line of research in Mexico and Latin America, which today we call "sociohistorical disaster research" (Rojas et al., 1987). Another publication, primarily authored by linguistic specialists, was titled *A reading of the quake in the capital's news media*, reflecting, in the words of Hans Sattele (who wrote the book's foreword), what he qualified as a "babelic" moment that the Mexican press experienced after the earthquake, offering a reading from the position of the affected subjects (Carbó et al., 1987). Yet another of these books, the product of a short-term research project directed by Ludka de Gortari and Juan Briseño and conducted in two earthquake-affected neighborhoods in the center of Mexico City, focused on a specific issue: the forms of cooperation and organization, with

an emphasis on attachment to neighborhood identity and political culture and solidarity, as an old survival strategy. Solidarity has always being present in Mexico
in the face of extraordinary events (Briseño & De Gortari, 1987). The last of this
four-book collection is a compilation of various short-term studies conducted sep-
arately by anthropologists, geographers, and historians who came together with
central conclusions including

> the general feeling among the population concerning the inefficacy of deci-
> sions made by the government, the lack of credibility and legitimacy of these
> decisions, which were further discredited by cases of fraud evinced in the
> construction of public buildings and housing complexes that were demol-
> ished by the quakes.
>
> (Di Pardo et al., 1987: p. 1).

Accompanying these efforts were other publications which also resulted from
short-term, focused research projects, including several that privileged an anthropological approach. Such is the case with one of the articles included in the
special issue of *Studies in Third World Societies* edited by Oliver-Smith in 1986 –
written by a group lead by Scott Robinson, an American anthropologist who had
many students in Mexican institutions, including myself – which emphazises the
relation between eathquake disasters and political hegemony (Robinson et al.,
1986). A second example is the Issue Number 48 of *Revista Mexicana de Sociología*, which was titled "Quake: disaster and society in Mexico City," published
the same year.[4] The third case refers to the First Paul Kirchhoff Colloquium,
conducted in April 1986 and published two years later under the title *Ethnology: Themes and Tendencies*, which included a supplement titled "Ethnology
of Disaster", featuring 11 articles authored by two dozen ethnologists and social
anthropologists.[5] The majority of the texts corresponded to specific studies based
on good ethnographies related to the earthquake: testimonials, shelters, and survivors. All these studies highlight an emphasis on context, that is, on prevalent
socioeconomic conditions as determinants of the disaster and the quake as the
detonator of the existing contradictions, implicitly recognizing that the active
agent was not the quake, but the increasing vulnerability, differently accumulated, as well as the population's unequal capacity to recover (Instituto de Investigaciones Antropológicas, 1988). The publication's authors did not continue
working strictly within this topic; had they done so, they would have subscribed
to the growing alternative current that was already developing and to which I will
refer to later on.

Like the aforementioned authors, many institutions became interested in the
issue of earthquakes and disasters in general, but CIESAS is one of the few (if not
the only) that has maintained such an interest from the perspective of the social sciences.[6] Because of that, and because of the quality of its academic products, there
is today a national and international recognition of the institution as a pioneer,
particularly in Latin America, in the development of the socio-anthropological
study of risk and disasters.[7]

Following the four already mentioned publications, some of us continued to work on the elaboration of historical catalogs that account for the presence and effects of natural hazards in Mexico. We have done this on the topic of earthquakes (García-Acosta & Suárez, 1996), followed by a second volume taking into account what historians and anthropologists scrutinize on the basis of that information (García-Acosta, 2001). With a similar methodology and covering a long period that begins from the pre-Hispanic period and ends in the 20th century, we put together a new catalog with information obtained from what we called "agricultural disasters" in Mexico, that is, disasters associated with hydrometeorological hazards: hurricanes, hail storms, droughts, and their impacts and effects (García-Acosta et al., 2003; Escobar Ohmstede, 2004).[8]

We already had an enormous amount of compiled information as well as perspectives of specialists from different parts of the world, leading us to identify in the topic a multidisciplinary potential unseen until then. Moreover, being researchers and not merely compilers of data, it was imperative to analyze this information in order to understand what had occurred and how such data could be useful for disaster research and disaster reduction. We dedicated ourselves to scrutinizing the existing literature for the types of analyses that had been conducted from a social science perspective, specifically, from anthropological and historical perspectives. I call this phase "from the catalogs to reflection".

Specialized literature coming from the social sciences was, in general, related to disasters conceived as natural; therefore, the natural hazard was the center of interest. At a global scale, anthropological interest began in Great Britain, followed by the United States in the 1950s, generally beginning from case studies also instigated by a natural hazard. Paradigmatic examples include the Lamington eruption, the Worcester tornado, and the Yap typhoons; these were pioneering studies carried out by anthropologists through the application of methods and techniques of anthropological work in the face of unusual conditions in the communities they studied. Little by little, they realized that these conditions were not so unusual, especially following the more thorough examination of the topic by William Torry (1979) and, particularly, Anthony Oliver-Smith.[9]

The greatest development had been achieved by the structural-functionalist school of sociology, especially after the 1960s, which proposed a theoretical focus on what came to be generically called "Sociology of Disasters". A great number of monographs, articles, and projects focused specifically on collective behavior, organizational analysis, and disaster response. The emphasis was on identifying the natural phenomenon as the active agent in the occurrence of disasters. Two of the most prominent examples of that epoch were Enrico Quarantelli (1925–2017) and Russell R. Dynes (1923–2019) of the prestigious Disaster Research Center.

Unsatisfied with the existing proposals, especially those made by North American sociology with a vision from the global north, and driven by the proposals identified as "alternative", which were derived from the research conducted by geographers and anthropologists (namely, Kenneth Hewitt, Ben Wisner, William Torry, and particularly Anthony Oliver-Smith) alongside various Latin American researchers, in the 1990s, we decided to promote a perspective related to vulnerability and prevention, inspired by a phrase that summarized our position in generic terms: disasters are NOT natural. In 1992, as I mentioned before, we founded LA RED, and the network attained widespread notoriety both because of its many relevant publications and because it covered a new topic that was particularly provocative within the framework of the United Nations' International Decade for Disaster Reduction (UNISDR). The execution of short-term research projects that were comparative in different countries and especially in Latin American regions inspired the phrase that was chosen as the title of LA RED's first publication which, in short time, placed

> at the disposition of researchers, not only within but also outside the region, invaluable reference material, to this day unmatched and difficult to obtain and, in that manner, promoted new efforts in the social study of disasters [. . .] contributing to the overcoming of isolation and achieving the feedback of ideas, which is the basis of any significant social research process.
>
> (Maskrey, 1993: p. XIII)

From its beginning, LA RED featured the participation of three anthropologists (Eduardo Franco, Anthony Oliver-Smith, and Virginia García-Acosta) as well as a geographer–quasi-anthropologist, Jesús Manuel Macías.[10] Even then, the proposal of a new line of research was maturing: the historical-anthropological study of disasters (García-Acosta, 1993).[11]

Following 1985, it was once again the presence of natural hazards in Mexico, in this case hurricanes, which sparked important reflections concerning their association with disaster: hurricane Paulina in 1997 and hurricane Isidore in 2002. These were typical visits received by the residents of the Mexican Pacific, Gulf of Mexico, and Caribbean coasts on a yearly basis but,[12] in this case, they triggered disasters that were the product of silent risks that were socially constructed and the product of accumulated vulnerabilities.

Pauline, first as a tropical storm and then as a category 4 hurricane in the Saffir-Simpson scale, moved through the Pacific coast, primarily affecting the states of Oaxaca and Guerrero, two of the four Mexican states with the highest levels of marginalization at the national level. As a disaster, it was the focus of local and regional analyses, among which stand out those conducted with extensive field-work of the highest anthropological level, made public years later (Vera, 2005a; Villegas, 2005),[13] as well as the relevant contributions of geographers.[14] It was one of the occasions when concepts like the social construction of disasters, and accumulated and differentiated vulnerabilities, demonstrated their efficacy in the analysis of disasters.

For its part, Isidore made landfall in the Yucatán Peninsula in 2002 and featured similar characteristics that were analyzed from an anthropological perspective. In two issues of the journal *Revista de la Universidad Autónoma de Yucatán*, with an analogous initiative to the one of CIESAS in 1985, the *Revista* sent a widespread call to not only to professors and researchers from the academic units of the Universidad Autónoma de Yucatán (UADY), but also to other research centers of the state. The response to this call was very generous and, a few months after the disaster, those two issues came out. The reflections from the perspective of anthropology and history came from a more mature framework, a matter of lessons that historical and contemporary reality insisted on demonstrating: hurricanes and disasters are not synonymous, nor they should be (Krotz, 2002; García-Acosta, 2002).

What has been the methodology followed in the conducted analyses? As we have said on other occasions, methodology constitutes the researcher's compass, and the one chosen must correspond with a certain theoretical model. The methods utilized in the study of disasters from an anthropological perspective, understanding disasters as processes, have been the product of the combination of long-term studies, comparison, and synchronic and diachronic perspectives within the framework of cultural and historical ecology.[15]

What does this anthropological perspective show us with empirical historical, contemporary evidence, combining documentary work, and fieldwork? It shows that risks and disasters related to natural hazards constitute multidimensional and multifactorial processes that result in the association of the hazard with a context of vulnerability. Disasters are processes that result from critical preexisting conditions within which vulnerability and, specifically, the ongoing and persistent social construction of risk, occupy a determining position in association with specific natural hazards. Therefore, it shows that disasters are not natural.

In this way, the call in one of the four publications that resulted from the CIE-SAS 1985 initiative that "This first attempt must, without a doubt, be expanded in the future" has become a reality and has yielded valuable fruit that has extended beyond strictly Mexican spaces. I hope that something similar occurs from the efforts initiated in the academy in response to the earthquake of 19 September 2017, the same day of the same month 32 years later. These efforts must be added to those that already exist in an effort to reduce risk and disaster.[16]

Anthropology and history: an unavoidable link[17]

Global history is littered with examples that show the non-natural condition of disasters, that natural hazards and disasters are not synonymous nor should they be, that natural hazards are also socially constructed and become socionatural hazards. These cases clearly show that disasters are the product of the accumulation of risks and vulnerabilities related to, but also derived from, the type of society and economy that has developed over time, and not from the increasing presence in frequency and magnitude of threats of natural origin.

In Mexico, a good example of the growing recognition of sociohistorical analysis of disasters is the work of Richard E. Boyer on Mexico City's 17th century floods, published in the 1970s. Boyer's analysis is not limited to the effects and impacts of the floods, but it carefully studies the context in which they occurred and clearly shows that disasters were the product of the confluence of natural, socioeconomic, political, and cultural factors of such a magnitude that the city remained flooded during nine long years: from 1629 to 1638 (Boyer, 1975).

Recognizing that disasters constitute processes means, as I mentioned, accepting the need to apprehend them from a diachronic point of view. Historical studies on disasters based on primary documentation,[18] carried out with consistency in Mexico and in some other countries of Latin America and the Caribbean,[19] but also in Europe and Asia,[20] have led to important conclusions. Mainly, that in the long term it is possible to identify how the progressive vulnerability and the accumulated vulnerabilities throughout the years and even centuries, as well as the permanent and continuous construction of risks, have been responsible for disasters. These disasters are associated with the presence of certain natural hazards both of a sudden impact (earthquakes, volcanic eruptions, tsunamis, hurricanes or tropical cyclones, hailstorms, frosts) and of slow impact (droughts, epidemics, pests). These studies show that the history of disasters is not a recounting of memorable events from the past, but rather a description of the ways in which this gradual and generalized process has tended to increase the vulnerability of the population starting from the invasion, conquest, and colonization.

The fruitful dialog between the social sciences and history, especially between anthropology and history, whose importance I have highlighted during the last three decades, is a dialog which follows the teachings of Fernand Braudel, Edward Thompson, and Enrique Florescano. It has become more and more necessary and, as I have tried to demonstrate, inescapable. It allows us to determine concepts, identify problems, emphasize specific elements, and, with all this, advance in the perception and comprehension that disasters are historical processes.

I will refer to two cases that, when examining the specific strategies developed during hurricanes, exemplify the aforementioned. The works of Konrad (1985, 1996) among pre-Hispanic and contemporary Maya, to which I already referred, dating back more than three decades, and one recently published by Raymundo Padilla (2018), a young Mexican scholar, historian, and anthropologist. Padilla's article, entitled "Prácticas históricas de alertamiento y protección ante huracanes en Baja California Sur" ("Historical warning and protection practices against hurricanes in Baja California Sur") is a product of working simultaneously with historical documents and fieldwork, as Konrad himself did in his time. In addition to reviewing the use of seemingly similar concepts such as practices, strategies, coping mechanisms, and those used by historians and anthropologists specialized in the study of disasters, Padilla emphasizes how much disasters, as well as practices developed in the particular case of recurrent hazards, are part of historical processes.

Documenting disasters as events located in specific times and places help to weave stories, but these stories need the *longue durée* perspective in order to really understand and apprehend what has happened (García-Acosta, 1996: pp. 17–18), as well as the comparative perspective that may show what Steward called "cultural patterns [. . .] which may develop in different parts of the world [and show how] cultural change is induced by adaptation to environment" (Steward, 1955: pp. 4–5). This is what Steward named cultural ecology, emphasizing "that the field was concerned with cultural adaptations to environment and the range of choices available to cultural groups" (Sutton & Anderson, 2018). Following him, our interest has focused on the study of the relationships between culture and environment envisioned through disaster research.

Exploring the past with a comparative and *longue durée* perspective reveals "a vast wealth of knowledge in relation to 'dealing with disaster', including 'adapting to change'". (Mercer, 2010: p. 248). Research has demonstrated the extraordinary role that history has played in understanding the build up of contemporary vulnerability, as well as in identifying traditional and ancient response mechanisms, and longstanding experiences in designing contemporary disaster risk reduction policies (García-Acosta, 2017).

Mexico in the global discussion of an "adjectival discipline": the Anthropology of Disasters

Can we name it thus? An adjectival discipline? Alternatively, is it a matter of a sub discipline, a branch of anthropology?

In the Introduction of one of the three books I consider paradigmatic in the anthropological study of disasters, *Catastrophe & Culture: The Anthropology of Disaster* (Hoffman & Oliver-Smith, 2002),[21] Oliver-Smith and Hoffman added as a subtitle the phrase "Why Anthropologists Should Study Disasters?" As time has gone by, anthropologists have learned, as these two outstanding academics pointed out, that the discipline has had important contributions to make to their study "and that disasters in turn have great expository relevance to the inquiries of their field [. . .] the alarming increase of disasters and their aftermaths, have clearly demonstrated how much light catastrophes can shed on the content of anthropological purview" (Oliver-Smith & Hoffman, 2002: p. 5).

Starting from the aforementioned concerns, it is natural to suppose that the school of cultural ecology, derived from the multilinear evolutionist approach, could have provided the main anthropological methodology in the field of disasters (Steward, 1955).[22] However, this was not the case, at least not explicitly, until several decades later. Adaptation has been one of the central concepts of anthropology since its inception in the 19th century. During that time, it was particularly interested in human biological and cultural evolution, an interest adopted by "anthropologists/disaster-experts" through the school of cultural ecology. They were (or are?) interested in the sociocultural adaptation processes in the society-nature relationship, in this case in relation to disasters associated with natural or socio natural hazards. As Oliver-Smith (2013, p. 276) said during the speech he gave when receiving the well-deserved 2013 Malinowski Award from the Society for Applied Anthropology:

> My approach to these issues comes from ecological anthropology and political ecology, which assert that society and nature are each involved in the shaping of the other, such that social features are expressed or instantiated ecologically, and environmental features are expressed socially in terms of exposure, vulnerability, resilience, or impacts.

During the 1970–1980s, anthropology maintained a constant interest in the subject, to the degree that it was possible to define an Anthropology of Disasters towards the end of that century.[23] In Mexico, as we saw before, it began during the middle of the 1980s. At the same time, the framework was refined to include political ecology (Oliver-Smith, 2002); this was influenced by a decisive focus on research in the global south "because vulnerability was most intense in the developing world, where most anthropologists worked," combining diachronic and comparative analyses, in and through the perspective of the *longue durée.* Anthropologists "and other social scientists began to draw an association between disasters and development, or under development. The enormous disparities in deaths and losses between disasters in the developed and developing worlds thus identified high disaster mortality and destruction as features of underdevelopment. This is, in essence, a political ecology of disasters" (Oliver-Smith, 2015: p. 547).

The global discussion about what I call here an adjectival discipline, about that branch of anthropology, continued to mature and advance, and Mexico has been permanently present in it. Mexico is present in two of the three publications I consider emblematic because they are the first of their time to present the anthropological and historical gaze (Oliver-Smith & Hoffman, 1999; Giordano & Boscovoinik, 2002). The same has occurred in various conferences that Latin American anthropologists celebrate periodically, such as those of the Brazilian Anthropological Association (ABA),[24] The Latin American Anthropological Association (ALA),[25] the Mexican Congress of Social Anthropology and Ethnology (COMASE),[26] and the Mexican-Brazilian Anthropological Meetings (EMBRA).[27] Special mention is due to the creation of the Society for Applied Anthropology of the Risk & Disaster Topical Interest Group (R&D TIG) which, since 2014, has been a space to convene specialist anthropologists from various parts of the world to discuss advances in this branch of Anthropology. Mexico, as a region researched, has been almost always present in its annual sessions, even though not always by Mexican scholars.[28]

While we can find specific anthropological studies that connect disasters with heritage, knowledges, reconstruction, or public policy, there are some compilations that offer Mexican cases concerning two particular themes: resettlement of populations at risk, usually post-disaster (Vera, 2009; Macías, 2001), and expressions of social vulnerability in a particularly sensitive area of the country, the southeast (Soares et al., 2014). Both offer compelling reflections coming mainly from anthropology on two of the core concepts in disaster studies: risk and vulnerability. They describe, in some cases specifically, decision-makers who often confuse "risk" and "vulnerability", which leads to improper, inadequate, and inefficient decision-making.

I must also highlight two recently published books, *Les catastrophes et l'interdisciplinarité* (García-Acosta & Musset, 2017) and *Governing Affect: Neoliberalism and Disaster Reconstruction* (Barrios, 2017). The first, recognizing the progress made by various disciplines within the social sciences, offers an interdisciplinary dialog, both conceptual and methodological, which involves five anthropologists, one of which offers a study about Mexico related to malaria. The imaginative title of the book by Roberto E. Barrios is an excellent synopsis of the intensive ethnographic work of the author, who studied the aftermath of four American cases, one of them in the Mexican state of Chiapas after the Grijalva River landslide. Barrios examines the role emotions (terror, disgust, or sentimental attachment) play in the effectiveness of governmental recovery, something new and powerful in the Anthropology of Disasters.

The Anthropology of Disasters in Mexico has traveled much of the narrated road in the hands of specialized geographers, with whom we share perspectives and methods. That is the case of Jesús Manuel Macías, a CIESAS researcher to whom I referred earlier as a geographer–quasi-anthropologist, who has gathered a team of mostly geographers been formed in projects leaded by him. He has done research for decades: since the case of gasoline explosions in the Guadalajara sewer system in 1992 (Padilla & Macías, 1993; Macías & Calderón, 1994), but mainly resettlement cases after disasters and, more recently, the case of tornadoes (in Michoacan and Mexico City) and their possible association with disasters (Macías, 2001, 2016).

The attention dedicated to one of these areas of disaster studies, that of relocations and resettlements as a result of disasters, but also due to forced migration or dam construction, has long attracted the attention of anthropologists beginning with Oliver-Smith himself. A good amount of research published mainly in CIESAS, with specific case studies in which have participated anthropologists and geographers, some professionals and consolidated others still in formation, realize what is frequently stated: there are very few successful cases[29] (Macías, 2001, 2009; Vera, 2009).

The conditions of vulnerability and risk of the population settled in the vicinity and on the slopes of two active volcanoes, Popocatépetl and Volcán de Colima, have been particularly attended by research teams in which anthropologists, geographers, and communicologists have actively participated in two directions.[30] On the one hand, analyzing and documenting in detail, based on case studies and fieldwork, local risk conditions, emergency plans, warning systems, and population evacuation or resettlement which, as I mentioned, have been generally unfortunate (Macías, 1999, 2005). On the other hand, and here anthropological participation has been decisive, there are those who conduct studies from a more cultural perspective, related to local perceptions of risk, cults, myths, legends, and ceremonies, in sum, participating in research that recuperates ancestral knowledge of the worldview associated with the behavior of volcanoes (Glockner, 1996a; Vera, 2005b). These studies, mainly done in the Popocatépetl risk area, have demonstrated the absolute need to recognize the importance of local beliefs and ideologies and, as such, work together with the local rural population in the design and implementation of prevention and evacuation plans (Glockner, 1996b: p. 162). Otherwise the failures in these actions, as we mentioned at the beginning of this chapter related to the 1982 Chichonal volcano eruption,[31] will continue to accumulate.

I hope I am not leaving out in this tight summary important reflections from an anthropological perspective. Just to finish this section I would like to mention an area which has been close to the Anthropology of Disasters: studies on epidemics and pests and their effects in the history of Mexico. Some closest to the field I have been describing are several projects in which has participated CIESAS ethnohistorian América Molina, who has come a long way in this field with resulting publications that cover two decades of fruitful research on epidemics (Molina, 1996, 2014; Molina et al., 2013).[32] Also within that field I should mention El Colegio de Michoacán historian Luis Arrioja, a scholar who has not only published several studies on the subject, but who also forms part of the group that was created in 2015 on both sides of the Atlantic, overtaking the Mexican borders to expand to the rest of the Latin American region, to carry out comparative studies with a historical-anthropological perspective of risk and disasters.[33]

From disaster to risk: new approaches and concepts

The identification of natural hazards as triggers of socially constructed processes drove us to change from an emphasis on knowing, studying, and monitoring disasters, towards an analysis of disaster as such. Nevertheless, recognizing the complexity of such processes, thanks to their study from historical and contemporary perspectives, and identifying them as the product of the accumulation of risks and vulnerabilities related to (but also derived from) the kind of society and economy that has developed over time and at a global scale,[34] we identified that the emphasis of our inquiries had to focus precisely on risk and vulnerability. That is, on growing risk and vulnerability associated with disasters and natural hazards. It is a matter of a paradigm shift that has been already documented (Gellert, 2012; Oliver-Smith, 2015). How are risk and vulnerability constructed? How have they been augmented? With what variables and denominators do they enter into relationships?

Recognizing the importance of adopting a focus on vulnerability and, later on, shifting to the identification of risk and its social construction as the central element in the occurrence of disasters, has been critical in the advance of visions towards disaster risk reduction. This remains far from identifying with a grotesque synonymy disaster risk reduction with climate change adaptation, whose differences and scant similitudes Mercer (2010) as well as the rich chapters included in the *Handbook of Disaster Risk Reduction Including Climate Change Adaptation* (Kelman et al., 2017) have clearly highlighted. Oliver-Smith asks the key question: "In climate change, are we talking about disasters? [. . .] climate change policy generally does not fully engage the issue of systemically imposed vulnerability, that is, the socially constructed outcomes of the way resources, wealth, and security are distributed in a society" (2013: pp. 278–279).

That transition from emphasizing disaster to emphasizing risk came accompanied by a series of concepts that, as abstractions of reality, allow a better understanding of the process as a whole. One of them, specifically that of the social construction of risk, calls attention precisely to a historical construction by identifying disasters as processes; it has demonstrated its analytical value, despite the fact that it has been given diverse meanings, which has led to confusion in its use, a matter I discuss in other publications (García-Acosta, 2005).[35]

The social construction of disaster risk is related to the generation and creation of vulnerable conditions and socioeconomic inequities, much like the production of new hazards that have been identified as socio natural hazards. They take the form of natural hazards and, in fact, are constructed upon elements of nature, but its materialization is the product of human intervention (Lavell, 2000). In sum, it is derived from a growing and accumulated material construction of disaster risk. The latter, which is defined as the probability of suffering future damages and losses, is frequently constructed upon other manifestations of risk and constitutes a latent condition that is frequently predictable. Such predictability is determined by the possibility of the presence of hazards and vulnerability.

Some associated concepts have helped to understand how social construction of risk generates and works. Like *extensive risk*, which refers to minor but recurrent risks, "associated with localized, mainly weather-related hazards with short return periods". Four underlying risk drivers characterize the accumulation of *extensive risks*: badly planned and managed urban development, the decline of regulatory ecosystem services, low-income households with no possibilities of participating in the formal market, and weak governance in cities and regions. "As such, *extensive risk* is endogenous to and produced by urban and economic development" (UNISDR, 2013: p. 68).

If the social construction of risk crystalizes and appears to become more visible in certain moments based on certain events, it is evident because of the fact that disasters and the very social construction of risk as such are themselves processes. By understanding them and studying them from a historical perspective, the emphasis is not on the isolated historical event, but instead on the search to identify, as Eric R. Wolf reiterates, the processes that lie behind the events. It is of the essence to, as we noted earlier, historicize disasters: to study them following Braudel's model of the *longue durée*, with its different components. Examining disasters over the course of the construction of risks that increase vulnerability, and which provoke a rise in the harmful effects of natural hazards and, therefore, the magnitude of disastrous events (García-Acosta, 2005).

At the same time, it was necessary to define the core concept that led to the alternative perspective in disaster analyses, which even changed the basis on which disasters were understood: vulnerability.[36] The anthropologist-historian Rogelio Altez has eloquently demonstrated its "historicity, that is to say, that which allows us to observe it as a socially produced condition [. . .] understand it as a historical product.[37] Moreover, as well as try to deconstruct it. Thus began to emerge those I have called the "vulnerability circumstances", side by side with the already classic "global vulnerability" coined by Gustavo Wilches-Chaux recognizing that vulnerability is not one or only and that has different dimensions. Among those "circumstances" are the following.[38]

- "Differentiated vulnerability": refers to that presented by social groups or communities in the presence of certain natural hazards, as it is not the same to be vulnerable to hurricanes or earthquakes, droughts or floods, and leads us to make the difference between vulnerability to hurricanes or vulnerability to volcanic eruptions.

- "Differential vulnerability": refers to the fact that not all people or social groups are equally vulnerable in the presence of natural or socio natural hazards and the occurrence of disasters due to social class, gender, age or ethnic group constitute some of the variables best identified through their strong association with vulnerability to disasters.
- "Progressive vulnerability": the accumulation of vulnerability's components throughout time, especially evident when studying cases over prolonged periods, as in the *longue durée*.[39]

Having worked for a long time around the concepts of vulnerability and social construction of risks, and having identified practices and actions to cope with natural hazards developed by communities in the present and past, we began to explore new concepts. Notions like adaptation and resilience, social capital, and social cohesion seemed to be useful (García-Acosta, 2009, 2018d); nevertheless, we felt that an inclusive concept was needed to give account of processes that are developed over time to cope with natural hazards. That is how the concept of social construction of disaster prevention was born parallel to the social construction of disaster risk, trying to look at (as I have said several times) the other side of the coin. This is the reason why I have linked it to the Greek god Janus: the god of the two sides, of two faces (García-Acosta, 2014).

As always happens in the course of the generation of conceptual proposals, this one was based on the search and identification of cases and examples coming from historical and contemporary reality. They came out of specialized literature, from different regions of the world, and particularly from experiences generated from Mexico. Given the purpose of this chapter, I will highlight the ones coming from anthropological research addressing Mexican examples.[40]

We did it by starting with the following hypothesis: societies are not and have never been passive entities in the face of natural hazards. They have formulated historical, social, and cultural ways to deal with potential risks and disasters, which are part of their culturally constructed ancestral and vernacular knowledge. The processes by which they have emerged are unfailingly resulting of the conditions in which each society grows and improves, that is to say that they are the product of a certain culture. So communities all around the world have imagined, created, constructed, rejected, and returned to imagine, create, and construct diverse practices, actions, and strategies that allow them to face and, above all, to prevent (many times really successfully) the effects related mainly to the imminent presence of a natural hazard (García-Acosta et al., 2012: p. 4).

With that hypothesis as a guide, the main objective has been to recuperate that culturally developed knowledge, primarily the processes by which it emerged. Cases of communities located in regions historically exposed to the presence of recurrent natural hazards, where can be identified elements of material culture from which have developed capacities and resilient strategies. I am referring to practices and actions, objectives, strategies, and capacities, usually collective and identified mostly at the local level, who have been managed to reduce risks associated with extreme events and that, in some cases, have become stronger through generations. What Faas (2016: p. 24) expresses as "an endorsement of enduring the value of historically ethnographic approaches that uncover the historical production of disaster [. . .] without rendering people as passive victims, but active agents capable of maneuver". At the end, the concept of social construction of prevention, from the local community perspective, refers to the methods by which the society develops preventive strategies against the recurrence of natural hazards, which causes permanent changes within the material culture and the organization of communities and groups affected. Never forgetting that

> adaptive capacity is not uniform throughout a community, but inflected by social, economic, and political factors. Each case will involve a complex interaction between local factors, such as knowledge, resources, power, adaptive capacity, and cultural values, and the intensity of an environmental threat.
>
> (Oliver-Smith, 2017: p. 214)

When I am talking about the social construction of prevention, and one of its expressions, adaptive strategies, I am following the way anthropology generally refers to adaptation, as "changes in belief and/or behavior in response to altered circumstances to improve the conditions of existence, including a culturally meaningful life" (Oliver-Smith, 2017: p. 208). I do agree with Oliver-Smith when he states that "an adaptation is part of a lifeway", and that we should not confuse it with coping, as "coping behavior involves immediate problem-solving and decision-making, including improvisation and creativity. Adaptations, on the other hand, are part of the fund of general knowledge and practice in a culture". In the same way the difference between coping strategies and adaptive strategies must be delineated, in that "coping strategies that effectively address challenges without longer-term negative outcomes may be adopted as established practices and come to constitute culturally sanctioned and socially enacted adaptations" (Oliver-Smith, 2017: p. 209).

The state of the art in Mexican anthropology of risk and disasters

Although there is still no wide recognition at a national level, Mexican Anthropology has had important advances and contributions to the Anthropology of Disasters. Let us explore some pathways to the future as a promising subdiscipline.

As I mentioned before, the conception and development of this last concept, the social construction of prevention accompanied by resilience and adaptive strategies, has been nourished by concrete case studies. In what follows, I will focus on those that come from historical and contemporary Mexico. Evidently, the catalyst was Konrad and his reflections about hurricanes among the Maya in the 1980s. In addition to what this Canadian anthropologist calls Maya strategies of survival in the face of hurricanes, which are first called "strategies of coping" (1985) and later substitutes for adapting strategies (1996), Konrad mentions the following: intensive agriculture in high fields, swidden agriculture, cultivation in terraces, hydraulic systems, backyard vegetable growing, and technical forestry adapted to the tropical forest. Using "Pre-Hispanic, colonial and contemporary texts as well as climatic data from the Caribbean region" he suggests that, "effective adaptation to the ecological effects of tropical storms helped determine the success of pre-Hispanic Maya subsistence strategies" (Konrad, 1996: p. 99). As we have asserted, successful adaptive strategies are built upon the organizational structures of the community based on the evaluation of events on the part of the affected group (García-Acosta, 2009).

During the last decade and a half, we can add to what we could call Konrad's invitation to carry out studies related to those type of strategies, based on empirical research coming out from historical documents as well as from ethnographic fieldwork. They have been as much the product of BA, master's, and doctoral theses, as of focused research projects. All refer to knowledge accumulated at the local community level, although they reflect diverse experiences, the majority of which could be classified as successful adaptations. Although I do not have space to mention them all, in what follows I will highlight those that I consider the most representative:

a)   Strategies, practices, and specific locally constructed actions:

- in face of floods related to the presence of El Niño in Veracruz: the cases of Tlacotalpan and Cosamaloapan (Angulo, 2006);
- spatial strategies designed throughout centuries by the inhabitants and authorities of the port of Campeche, in face of the recurrent presence of tropical cyclones and other climatic events (Cuevas, 2010, 2012a, 2012b);
- differential monitoring and warning strategies in the face of hurricanes as a chronic hazard, generated by inhabitants of communities located in the Mexican Pacific Coast (Baja California and Colima) (Padilla, 2014, 2018);
- those developed by the population of Chetumal, Quintana Roo, in the presence of recurring hurricanes; experiences translated into knowledge that have been articulated with technological advances and the actions promulgated by governmental institutions to protect, prepare, and recover, including mitigating impacts and attenuating damages (Rodríguez, 2018).
- strategies and preventive actions that resulted from the knowledge accumulated in the northeast of Sinaloa in the presence of recurrent floods associated with monsoon climate: the case of Guasave (Angulo, 2018)
- "good practices" in several communities in Yucatan Peninsula (Quintana Roo and Yucatan) developed by citizens, authorities, and social organizations in order to address hydrometeorological phenomena (hurricanes, heavy rains, strong winds) and learn to live with them, producing their own space to live safer (Cuevas, 2012b);
- local traditional practices relative to the environment, natural resources, and fishing practices associated with the presence of hydro meteorological

phenomena and climate change in the coastal zones of the Yucatan Peninsula (Rio Lagartos Biosphere Reserve) (Audefroy & Cabrera, 2012).

b) Related to technology, architecture, and adequacy to the surrounding environment:

- vernacular housing in different Mexican regions as a product of different processes to adapt to diverse weather and climate conditions, and also to extreme events, always incorporating local knowledge: cases coming from Baja California, Campeche, Chihuahua, Coahuila, Tabasco, Veracruz, and Yucatan. (Audefroy, 2012, 2015);
- technological adaptive responses in face of water scarcity (channels and *chultunes)* among Maya inhabitants in the Yucatan peninsula, that have been fundamental and original strategies to cope and recover from so long periods of drought (Audefroy & Cabrera, 2018).

c) Weather and climate knowledge and management related to traditional practices:

- climatic interpretations of C'holes ethnic group of northern Chiapas identified through phenological indicators and expressed through agricultural rituals, including practices that represent capacity adaptation bias (Briones, 2012);
- weather forecasting and its cultural expressions based on historical memory and ancestral knowledge of the environment which, in many cases, explains the success of traditional cultures (Albores & Broda, 1997; Toledo & Barrera-Bassols, 2008).

With the aforementioned cases almost all the coastal Mexican region is covered, as nearly all the examples are related to hydro meteorological hazards or events: Pacific coast (Baja California, Colima, Chiapas), Caribbean and Gulf of Mexico (Campeche, Quintana Roo, Tabasco, Yucatan), including two inland Mexican States, Chihuahua and Coahuila representing spaces with dry or desert climate.

Without a doubt, we will have to continue exploring examples like the ones aforementioned in order to compare and contrast as much in Mexico as with cases from other Latin American regions in particular, or the global south in general, privileging the gaze focused on the local communitarian. Continuing the study of adaptation strategies can provide us with clues to understand how social groups have used their collective imagination in order to improve their chances of survival. Perhaps we have not searched enough, or maybe not where we should have: namely in community dynamics, emphasizing more local than global perspectives to define the risky circumstances and thus reduce the effects and impacts that can generate the presence of a specific natural hazard. The key element to continue in this path of research is to identify the reasons why some communities manage to change their modes of action, while others reproduce the same perilous situations. We must identify when and how risk is socially constructed at different levels, and distinguish how through different mechanisms communities have collectively developed strategies, actions, and practices of prevention that have resulted in increasing resilience.

The contributions of anthropology to these issues have allowed us to identify variables that have little visibility and that are even ignored in risk management. Without a doubt, Mexico has reached an important development along the road we have taken, which I have tried to show in this chapter. However, we still require collective efforts to identify relevant gaps in our acquired knowledge and to attend to one of the major challenges, which is to translate it into public policies.

Notes

1 I am grateful to those colleagues who helped make this text more readable in English: Roberto E. Barrios (Dept. of Anthropology, Southern Illinois University Carbondale) and Rachel A. Cypher (PhD Candidate, Dept. of Anthropology, University of California, Santa Cruz).

2 CIESAS, "Mexico's first specialized research and postgraduate studies center on anthropology and history" (Krotz, 2018) nowadays a leader in the postgraduate education of Anthropology in Latin America, was founded as CISINAH (Center for Advanced Research of the National Institute of Anthropology and History) in 1973 by three eminent anthropologists: Gonzalo Aguirre Beltrán, Guillermo Bonfil, and Ángel Palerm, this last being its first director (García-Acosta, 2018a).

3 Casa Chata is the name of the 18th century emblematic building that houses CIESAS since 1973, and the publications have that name.

4 The majority of authors were sociologists and political scientists, except for an anthropologist (Cecilia Rabell) and an archaeologist (Linda Manzanilla). It is worthwhile to highlight here the recent publication of a special issue of the same magazine entitled "Earthquakes: 1985 and 2017", which involved two anthropologists: Raymundo Padilla and Mariana Mora (*Revista Mexicana de Sociología*, 2018).

5 Members of various institutions for research and studies in Anthropology: the ENAH (National School of Anthropology and History, the Institute of Anthropological Research, the Institute of Social Research of the UNAM (National Autonomous University of Mexico), and CIESAS itself.

6 Which is evidenced through by numerous research projects, whose results have been published in books and articles, as well as by dozens of undergraduate and graduate theses. Only in Mexico, graduate theses developed within CIESAS, or in collaboration with CIESAS researchers, have developed studies in virtually all the Mexican Republic, attending cases of a variety of natural hazards and disaster processes. Some of them are quoted in this chapter.

7 It is enough to check the bibliography that appears at the end of this chapter. A summary on the emergence and development of that line of research appeared in Valencio and Siena (2014: pp. 15–20).

8 Both volumes are available as ebooks since 2015, also published by the Fondo de Cultura Económica.

9 All this is mentioned more extensively in the Introduction to this volume.

10 In LA RED publications, which to date can be located all online (www.desenredando. org), the analysis and evaluation of Mexican cases were always present: cfr. Journal *Desastres & Sociedad* num. 1 & 4; *Los desastres no son naturales*, 1993 (ch. 5 & 8); *Al norte del Río Grande. Ciencias sociales y desastres desde una perspectiva norteamericana*, 1994 (ch. 4); *Viviendo en riesgo. Comunidades vulnerables y prevención de desastres en América Latina*, 1994 (ch. 15); *Estado, sociedad y gestión de los desastres en América Latina. En busca del paradigma perdido*, 1996 (ch. 6); *Desastres: Modelo para armar. Colección de piezas de un rompecabezas social*, 1996 (ch. 4, 6, 13, & 15); *Navegando entre brumas. La aplicación de los sistemas de información geográfica al análisis de riesgo*, 1998 (ch. 4 & 9); *Escudriñando* en los desastres a todas las escalas.

Concepción, metodología y análisis de desastres en América Latina utilizando DesInventar, 1999 (ch. 6 & 8); *Qu-ENOS pasa? Guía de LA RED para la gestión radical de riesgos asociados con el fenómeno ENOS*, 2007 (ch. Mexico) and its English version: *ENSO what? LA RED guide to getting radical with ENSO risks; ENOS Variabilidad climática y el riesgo de desastre en las Américas: procesos, patrones, gestión*, 2008 (ch. Mexico). In fact, one of LA RED's meetings of what we identified as "the ENSO Project", was held in Oaxaca, whose Unidad Estatal de Protección Civil (Oaxaca State Civil Protection Unit) adopted what they called "LA RED paradigm" and published in 2007 a booklet entitled *Oaxaca en LA RED. En busca de un paradigmo alternativo para la gestión de riesgo: la prevención* (*Oaxaca in LA RED. In search of an alternative paradigm for risk management: prevention*).

11 Evidenced, among others, in the three volumes of *Historia y desastres en América Latina* (*History and Disasters in Latin America:* García-Acosta, 1996, 1997, and 2008) by LA RED and CIESAS, in each of which appear at least three chapters on Mexican cases that cover from the pre-Hispanic period to the 20th century.

12 *Historia y memoria de los huracanes y otros episodios hidrometeorológicos en México* (*History and memory of Hurricanes and other Hydro-meteorological Episodes in Mexico*) will soon be published with information that covers from the pre-Hispanic era until the second half of the 20th century. The presence of hurricanes in Mexico dates back to immemorial time, with spatial and temporal patterns clearly defined. It includes four case studies of hurricanes that cross-cut Mexican history through the centuries (García-Acosta & Padilla, forthcoming).

13 The Project was funded by Georgia State University, as part of a research conducted by the psychologist Fran Norris and the anthropologist Arthur Murphy.

14 Among them are Jesús Manuel Macías, from CIESAS and Diana Liverman.

15 An extended version on these issues was published in Spanish in 2004 and later in English in 2015 (García-Acosta, 2004, 2015). In them I refer to four sources of inspiration: Emile Durkheim (1858–1917), Fernand Braudel (1902–1985), Ángel Palerm (1917–1980), and Eric Wolf (1923–1999), to whom I have to add the Mexican historian Enrique Florescano (1937–); they all, in different moments, have insisted that methodology is maybe the only way to unify social scientists.

16 An ongoing doctoral thesis in Anthropology by María Rodríguez in El Colegio de Michoacán, Mexico, is a focused, retrospective analysis in Jojutla, Morelos, one of the most affected municipios; its title is *El desastre de 2017 en Jojutla, Morelos: sismo, sociedad y políticas públicas* (*2017 disaster in Jojutla, Morelos: earthquake, society and public policy*).

17 A first version of these ideas appeared in García Acosta (2018b).

18 That means first-hand documents produced by those who lived the moment studied (official, ecclesiastical or private archives, chronicles, stories, writings of travelers, newspapers) supplemented with secondary sources.

19 The following examples can be mentioned: for Latin America in general, the three volumes of *History and Disasters in Latin America* aforementioned, which include case studies for the entire region over six centuries. For Mexico the works of Isabel Campos (recently deceased), América Molina, and Raymundo Padilla. In the case of Venezuela, there is the extensive work of the anthropologist and historian Rogelio Altez.

20 There are very complete studies, such as the compilation of Christof Mauch and Christian Pfister (2009), and specific studies by Pfister himself (Switzerland), Franz Mauelshagen and Gerrit Jasper Schenk (Germany), Armando Alberola (Spain), Christian Rohr (Austria) and, of course, the extensive work of Greg Bankoff and J. C. Gaillard on the Philippines.

21 The other two are Oliver-Smith & Hoffman, *The Angry Earth. Disaster in Anthropological Perspective*, and Giordano and Boscoboinik, *Constructing Risk, Threat, Catastrophe. Anthropological Perspectives*, published in 1999 and 2002 respectively. A new and revised version of *The Angry Earth* is ongoing.

22 Angel Palerm introduced the work of Julian H. Steward published in 1955, to Mexican anthropologists in the 1960s, but it had to be read in English. CIESAS with the Universidad Iberoamericana (Ibero-American University) and the Universidad Autónoma Metropolitana-Iztapalapa finally made it available in Spanish thanks to the Series "Classic and Contemporaneous Anthropology" in 2014: half a century later!

23 The three aforementioned compilations realize what can be called the "certificate of naturalization" by the Anthropology of Disasters. A brief account of the evolution of the discipline in the field appeared as an entry in the *International Encyclopedia of Anthropology* (García Acosta, 2018c).

24 At the ABA 30th Meeting, in 2016 (Round table "Anthropology and disasters: a comparative reading" and the Working Group "Anthropology of Catastrophes: Approaches and Perspectives").

25 In the ALA meetings in 2015 (Symposium "The Anthropology of Risk and Disaster: Latin American Exchanges with Global Perspectives and Local Cases") and 2017 (Worktable "Anthropology of Disasters in Latin America and the Caribbean: State of the art" and Symposium "Vulnerability and Risk: Anthropology of Disasters, the local and the global").

26 In the COMASE Meetings (named before CONAE: Congreso Nacional de Antropología Social y Etnología, National Congress of Social Anthropology and Ethnology) held in 2010 (Symposium "The Contributions of Anthropology to the Study of Risk and Disasters in Mexico") and 2012 (Symposium "Strategies, Risks and Natural Hazards: the Recovery of Social and Cultural Memory").

27 Corresponding to 2013 sessions (Working Group "Anthropology of Risk and Disasters: Cross-Sectional Views") and 2015 ones (Panel: "Socio-environmental Risks and Water Management").

28 Held in Albuquerque 2014, Pittsburgh 2015, Vancouver 2016, and Santa Fe 2017. Only in the one that occurred in Philadelphia 2018, although 120 risk and disaster panels and papers were presented, no paper on Mexico showed up. DICAN (Disaster and Crisis Anthropology Network) is another interesting network, born at the other side of the Atlantic as part of the EASA (European Association of Social Anthropologist), whose aim is to "facilitate contacts between anthropologists in order to enable exchange, communication and focused discussions about the anthropological contributions to the study of crises and disasters" (www.easaonline.org/networks/dican/), with a small participation of Latin American cases.

29 Macías (2001, 2009) and Vera (2009) compilations include Mexican cases in Chiapas, Puebla, Veracruz, and Yucatan related to disasters linked to natural hazards: hurricanes (Isidore in 2002 and Stan in 2005) or volcanic eruptions.

30 In the case of research in the so called Popocatépetl region, it is necessary to mention the role played by the Centro Universitario para la Prevención de Desastres Regionales (CUPREDER, University Center for the Prevention of Regional Disasters), which was created in 1995 in Puebla and since then has been under the leadership of the communicologist Aurelio Fernández. Up to now CUPREDER has developed, always in a close relation with geographers and anthropologists, multiple projects of incidence, in planning for emergencies' care and handling, about the relationship between disaster and environmental deterioration and the relationship of all these issues with the design of public policies (www.cupreder.buap.mx/).

31 A couple of postgraduate Theses in anthropology and some publications were produced later, related to the Chichonal eruption, but mainly with the conflictive and difficult migration processes that followed. See the following written by Zoque anthropologists: Domínguez (2018); Reyes (2007).

32 See also Paola Peniche's (2010) study on epidemics in Yucatan in the 18th century.

33 As a result of this collaboration, in which have participated scholars from other Mexican institutions (CIESAS, INAH, University of Colima, among others), as well as the

CONACYT Network "Interdisciplinary Studies on Vulnerability, Social Construction of Risk, Natural and Biological Hazards" (http://sociedadyriesgo.redtematica.mx/) we have several publications (i.e., Arrioja & Alberola, 2016; Alberola, 2017), to be enriched by investigations carried out within the project "Climate, Risk, Disaster and Crisis on Both Sides of the Atlantic during the Little Ice Age" (CRICATPEH: 2018–2020) funded by the Ministry of Economy of the Spanish Government.

34 Inevitable to recall here one of the famous phrases our dear colleague, geographer Allan Lavell, expressed to receive in 2015 the UNISDR Sasakawa Disaster Risk Reduction Award: "risk management should be part of the DNA of the development" (www.iucn.org/node/17691).

35 Some graduate thesis in Anthropology have dealt with this concept in different Mexican contexts. Among them are Briones (2008) researching in Oaxaca, Hernández (2006) in Guerrero, and Rodríguez (2007) in Baja California.

36 This concept, as A.J. Faas (2016) and Oliver-Smith (2017) well address, remains a complex problem, and is frequently viewed as a mere matter of exposure or as a descriptive concept that has been dehistoricized, subtracting its criticality.

37 His multiple publications allow documentation and clearly define the vulnerability to which he calls "analytical category". See Altez (2016, 2017).

38 A more finished version of these definitions appeared in García-Acosta (2018b).

39 There are other "vulnerabilities" identified by researchers in the disaster and risk field. One of them which I find particularlly revealing is that of "procedural vulnerability", which has not to do with the context and the people in face of natural hazards and disasters, but with the failure of stakeholders and policy-makers, where "the propensity of harm lies not simply in the end-point or context of particular hazards, but in the processes and assumptions that inform research questions, methods and outcomes in hazards research" (Veland et al., 2013: p. 314). "Procedural vulnerability" is also mentioned by Faas in his chapter on Ecuador included in this volume.

40 Among the experiences that nurtured these ideas are the following: the creation in 2010 and the results of the "Risk and Vulnerability Network: Social Strategies of Prevention and Adaptation" as the main outcome of an international project funded under the EU-Mexico cooperation by FONCICYT/CONACYT and based on CIESAS (http://redriesgoresiliencia.ciesas.edu.mx/), and the organization and papers presented in three consecutive conferences: COMASE 2010 and 2012, and ALA 2015, aforementioned.

References

Alberola, A. (ed.). (2017) *Riesgo, desastre y miedo en la península Ibérica y México durante la Edad Moderna*. Alicante and Zamora, Universidad de Alicante, El Colegio de Michoacán.

Albores, B. & J. Broda (coord.). (1997) *Graniceros: cosmovisión y meteorología indígenas en Mesoamérica*. Mexico City, Universidad Nacional Autónoma de México.

Altez, R. (2017) *Historia de la vulnerabilidad en Venezuela. Siglos XVI-XIX*. Sevilla, Universidad de Sevilla, Diputación de Sevilla.

Altez, R. (2016) Aportes para un entramado categorial en formación: vulnerabilidad, riesgo, amenaza, contextos vulnerables, coyunturas desastrosas. In: Arrioja, L. & Alberola, A. (eds.). *Clima, desastres y convulsiones sociales en España e Hispanoamérica, siglos XVII-XX*, Alicante and Zamora, Universidad de Alicante, El Colegio de Michoacán, pp. 21–40.

Angulo, F. (2018) Vulnerabilidad en el noroeste de Sinaloa por inundaciones asociadas al clima monzónico. Estrategias y acciones preventivas, caso de estudio: Guasave. PhD Thesis. Mexico City. Universidad Nacional Autónoma de México.

Angulo, F. (2006) *El Niño, inundaciones y estrategias adaptativas en Tlacotalpan y Cosamaloapan, Veracruz.* Masters Thesis. Mexico City, Universidad Autónoma de la Ciudad de México.

Arrioja, L. & Alberola, A. (eds.). (2016) *Clima, desastres y convulsiones sociales en España e Hispanoamérica, siglos XVII–XX,* Alicante, Zamora, Universidad de Alicante, El Colegio de Michoacán.

Audefroy, J. F. (2015) *La integración del conocimiento local para la adaptación al cambio climático en Veracruz.* [Paper] Simposio La antropología del riesgo y del desastre: intercambios latinoamericanos con miradas globales y casos locales, Mexico City, IV Congreso Latinoamericano de Antropología, Asociación Latinoamericana de Antropología.

Audefroy, J. F. (2012) Adaptación de la vivienda vernácula a los climas en México. In: García-Acosta, V., Audefroy, J. A. & Briones, F. (eds.). *Estrategias sociales de prevención y adaptación. Social Strategies for Prevention and Adaptation.* Mexico City, CIESAS, FONCICYT (CONACYT/European Community), pp. 95–108.

Audefroy, J. F. & Cabrera, B. N. (2018) La integración del conocimiento local para la adaptación al cambio climático: el caso de la costa yucateca. In: Audefroy, J. F. & Padilla Lozoya, R. (coord.) *Desastres asociados a Fenómenos Hidrometeorológicos.* Morelos, Mexico, Instituto Mexicano de Tecnología del Agua, pp. 173–201.

Audefroy, J. F. & Cabrera, B. N. (2012) Las sequías en el área maya: estrategias tecnológicas y adaptativas. In: García-Acosta, V., Audefroy, J. A. & Briones, F. (eds.). *Estrategias sociales de prevención y adaptación. Social Strategies for Prevention and Adaptation.* Mexico City, CIESAS, FONCICYT (CONACYT/European Community), pp. 116–126.

Báez-Jorge, F., Rivera, A. & Arrieta, P. (1985) *Cuando ardió el cielo y se quemó la tierra. Condiciones socioeconómicas y sanitarias de los pueblos zoques afectados por la erupción del volcán Chichonal.* Mexico, Instituto Nacional Indigenista.

Barrios, R. E. (2017) *Governing Affect: Neoliberalism and Disaster Reconstruction.* Nebraska, University of Nebraska Press.

Boyer, R. E. (1975) *La gran inundación, vida y sociedad en la ciudad de México (1629–1638).* Mexico, Secretaría de Educación Pública.

Braudel, F. (1958) La longue durée. *Annales Economies Societes Civilisations.* 4, pp. 725–753.

Briones, F. (2012) Saberes climáticos en la agricultura de los ch'oles de Chiapas. In: García-Acosta, V., Audefroy, J. A. & Briones, F. (eds.). *Estrategias sociales de prevención y adaptación. Social Strategies for Prevention and Adaptation.* Mexico City, CIESAS, FONCICYT (CONACYT/European Community), pp. 109–115.

Briones, F. (2008) *La construction sociale du risque: l'Isthme de Tehuantepec face au phénomène climatique 'El Niño' (Oaxaca, Mexique).* PhD Thesis in Anthropology. Paris, École des hautes études en sciences sociales.

Briseño, J. & De Gortari, L. (1987) *De la cama a la calle: sismos y organización popular.* Mexico, CIESAS.

Bueno Castellanos, C. (2010) Logros y vicisitudes de la Antropología Mexicana a través de su historia. *Revista Educacón Superior y Sociedad* (UNESCO). 15 (1), pp. 155–178.

Carbó, T., Franco, V., de la Torre, R. & Coronado, G. (1987) *Una lectura del sismo en la prensa capitalina.* Mexico, CIESAS.

Cuevas, J. (2012a) *Good Practices Used at the Peninsula de Yucatan, Mexico.* Background paper prepared for the 2013 Global Assessment Report on Disaster Risk Reduction. UNISDR, Geneva.

Cuevas, J. (2012b) Cuando el agua corre . . . Estrategias y prácticas espaciales para convivir con fenómenos hidrometeorológicos. El caso de la ciudad de Campeche, Mexico. In: García-Acosta, V., Audefroy, J. A. & Briones, F. (eds.). *Estrategias sociales de prevención y adaptación. Social Strategies for Prevention and Adaptation.* Mexico City, CIESAS, FONCICYT (CONACYT/European Community), pp. 127–136.

Cuevas, J. (2010) *Aquí no pasa nada. Estrategias y prácticas espacio-temporales para hacer frente a fenómenos hidrometeorológicos en la ciudad de San Francisco de Campeche.* Masters Thesis in Social Anthropology. Mexico City, CIESAS.

Di Pardo, R., Novelo, V., Rodríguez, M., Calvo, B., Galván, L. E. & Macías, J. M. (1987) *Terremoto y sociedad.* Mexico, CIESAS.

Domínguez, F. (2018) *Desplazamientos territoriales, flujos migratorios y erupciones volcánicas entre los zoques de Chapultenango, Chiapas: la constitución y reproducción de una diáspora indígena.* PhD in Social Anthropology, Mexico, Universidad Iberoamericana.

Escobar, A. (2004) *Desastres agrícolas en México. Catálogo histórico.* Vol. II: Siglo XIX (1822–1900). Mexico, Fondo de Cultura Económica, CIESAS.

Faas, A. J. (2016) Disaster Vulnerability in Anthropological Perspective. *Annals of Anthropological Practice.* 40 (1), pp. 14–27.

García-Acosta, V. (2018a) Angel Palerm. In: Callan, H. (ed.). *The International Encyclopedia of Anthropology.* New York, John Wiley & Sons, Ltd., pp. 4510–4515.

García-Acosta, V. (2018b) Vulnerabilidad y desastres. Génesis y alcances de una visión alternativa. In: González de la Rocha, M. & Saraví, G. A. (coords.). *Pobreza y Vulnerabilidad: debates y estudios contemporáneos en México.* Mexico, CIESAS, pp. 212–239.

García-Acosta, V. (2018c) Anthropology of Disasters. In: Callan, H. (ed.). *The International Encyclopedia of Anthropology.* New York, John Wiley & Sons, Ltd., pp. 1622–1629.

García-Acosta, V. (2018d) Cohesión social y reducción de riesgos de desastres. Otros conceptos a explorar. *Regions & Cohesion.* 8 (1), pp. 79–90.

García-Acosta, V. (2018e) Los sismos como detonadores. In: *Revista Rutas de Campo,* Mexico City, Instituto Nacional de Antropología e Historia, vol. 3, pp. 122–126.

García-Acosta, V. (2017) Building on the Past: Disaster Risk Reduction Including Climate Change Adaptation in the *Longue Durée.* In: Kelman, I., Mercer, J. & Gaillard, J. C. (eds.). *Handbook of Disaster Risk Reduction Including Climate Change Adaptation.* London, Routledge, pp. 203–213.

García-Acosta, V. (2015) Historical Perspectives in Risk and Disaster Anthropology: Methodological Approaches. In: Wisner, B., Gaillard, J. C. & Kelman, I. (eds.). *Disaster Risk. Critical Concepts in the Environment.* London and New York, Routledge, vol. 3, pp. 271–283.

García-Acosta, V. (2014) De la construction sociale du risque a la construction sociale de la prévention: les deux faces de Janus. In: Bréda, C., Chaplier, M., Hermesse, J. & Piccoli, E. (dir.). *Terres (dés) humanisées: ressources et climat,* Investigations d'Anthropologie Prospective 10, Laboratoire d'anthropologie prospective, Louvain, Academia-L'Harmattan Editions, pp. 297–318.

García-Acosta, V. (2009) Prevención de desastres, estrategias adaptativas y capital social. In: Koff, H. (ed.). *Social Cohesion in Europe and the Americas: Power, Time and Space.* Brussels, P.I.E.-Peter Lang Editions Scientifiques Internationales, pp. 115–130.

García-Acosta, V. (coord.). (2008) *Historia y desastres en América Latina,* vol. III. Mexico, LA RED-CIESAS.

García-Acosta, V. (2005) El riesgo como construcción social y la construcción social de riesgos. *Desacatos. Revista de Antropología Social.* 19, pp. 11–24.

García-Acosta, V. (2004) La perspectiva histórica en la Antropología del riesgo y del desastre. Acercamientos metodológicos. *Relaciones. Estudios de historia y sociedad.* XXV (97), pp. 123–142.

García-Acosta, V. (2002) Huracanes y/o desastres en Yucatán. *Revista de la Universidad Autónoma de Yucatán.* 17 (223), pp. 3–15.

García-Acosta, V. (2001) *Los sismos en la historia de México. v. II: El análisis social.* Mexico, Fondo de Cultura Económica, CIESAS, Universidad Nacional Autónoma de México.

García Acosta, V. (coord.). (1997) *Historia y desastres en América Latina,* vol. II. Lima, LA RED-CIESAS-Tercer Mundo Editores.

García Acosta, V. (coord.). (1996) *Historia y desastres en América Latina,* vol. I. Bogota, LA RED-CIESAS-Tercer Mundo Editores.

García-Acosta, V. (1993) Enfoques teóricos para el estudio histórico de los desastres naturales. In: Maskrey, A. (comp.). *Los desastres no son naturales.* Bogota, LA RED-Tercer Mundo Editores, pp. 155–166.

García-Acosta, V., Audefroy, J. F. & Briones, F. (eds.). (2012) *Estrategias sociales de prevención y adaptación. Social Strategies for Prevention and Adaptation.* Mexico City, CIESAS/FONCICYT (CONACYT/European Community).

García Acosta, V. & Musset, A. (dir.). (2017) *Les Catastrophes et l'interdisciplinarité: dialogues, regards croisés, pratiques.* Investigations d'Anthropologie Prospective, Laboratoire d'Anthropologie Prospective, Université catolique de Louvain. Louvain, Editorial Academia-L'Harmattan.

García-Acosta, V. & Padilla, R. (forthcoming) *Historia y memoria de los huracanes y otros episodios hidrometeorológicos en México. Cinco siglos: del año 5 Pedernal a Janet.* Mexico City and Colima, CIESAS, Universidad de Colima, Fondo de Cultura Económica.

García-Acosta, V., Pérez, J. M. & Molina, A. (2003) *Desastres agrícolas en México. Catálogo histórico. V. I: Épocas prehispánica y colonial (958–1822).* Mexico, Fondo de Cultura Económica, CIESAS.

García-Acosta, V. & Suárez, G. (1996) *Los sismos en la historia de México,* vol. I. Mexico, Fondo de Cultura Económica, CIESAS, Universidad Nacional Autónoma de México.

Gellert, G. I. (2012) El cambio de paradigma: de la atención de desastres a la gestión del riesgo. *Boletín Científico Sapiens Research.* 2 (1), pp. 13–17.

Giordano, C. & Boscoboinik, A. (eds.). (2002) *Constructing Risk, Threat, Catastrophe. Anthropological Perspectives.* Fribourg, Fribourg University Press.

Glockner, J. (1996a) *Los volcanes sagrados. Mitos ¿y rituals en el Popocatépetl y la Iztaccíhuatl.* Mexico, Grijalbo.

Glockner, J. (1996b) El sueño y el sismógrafo. *Desastres & Sociedad.* 6 (4), pp. 157–162.

Hernández, J. A. (2006) *La construcción social del riesgo a inundaciones y su asociación con El Niño. El caso de la subcuenca del río Omitlán, Guerrero. 1982–83 y 1997–98.* Masters Thesis in Social Anthropology. Mexico, CIESAS.

Instituto de Investigaciones Antropológicas. (1988) Etnología del desastre. In: *La Etnología: Temas y Tendencias. I Coloquio Paul Kirchhoff.* Mexico, Universidad Nacional Autónoma de México, pp. 229–362.

Kelman, I., Mercer, J. & Gaillard, J. C. (ed.). (2017) *Handbook of Disaster Risk Reduction Including Climate Change Adaptation.* London, Routledge.

Konrad, H. W. (1996) Caribbean Tropical Storms: Ecological Implications for Pre-Hispanic and Contemporary Maya Subsistence Practices on the Yucatán Peninsula. *Revista mexicana del Caribe.* I (1), pp. 98–130.

Konrad, H. W. (1985) Fallout of the War of the Chacs: The Impact of Hurricanes and Implications for Prehispanic Quintana Roo Maya Processes. In: Thompson, M., García, M. T. & Kense, F. J. (eds.). *Status, Structure and Stratification: Current Archaeological*

Reconstructions. Calgary, The University of Calgary Archaeological Association, pp. 321–330.

Krotz, E. (2018) Anthropology in Mexico. In: Callan, H. (ed.). *The International Encyclopedia of Anthropology*. New York, John Wiley & Sons, Ltd. DOI:10.1002/9781118924396.wbiea2115.

Krotz, E. (2002) Reflexiones desde la antropología sobre el huracán Isidore. *Revista de la Universidad Autónoma de Yucatán*. 17 (223), pp. 16–31.

Lavell, A. (2000) Desastres y Desarrollo: hacia un entendimiento de las formas de construcción social de un desastre. El caso del huracán Mitch en Centroamérica. In: Garita, N. & Nowalski, J. (eds.). *Del desastre al desarrollo humano sostenible en Centroamérica*. San José de Costa Rica, BID – Centro Internacional para el Desarrollo Humano Sostenible, pp. 7–45.

Macías, J. M. (coord.). (2016) *El tornado del Zócalo de la ciudad de México. la ocurrencia del evento tornádico del 1 de junio de 2012 en la ciudad de México y el área metropolitana*. Mexico, CIESAS.

Macías, J. M. (coord.). (2009) *Investigación evaluativa de reubicaciones humanas por desastre en México*. Mexico, CIESAS.

Macías, J. M. (coord.). (2005) *La disputa por el riesgo en el volcán Popocatépetl*. Mexico, CIESAS.

Macías, J. M. (2001a) *Descubriendo tornados en México. El caso del tornado de Tzintzuntzan*. Mexico, CIESAS.

Macías, J. M. (comp.). (2001b) *Reubicación de comunidades humanas. Entre la producción y la reducción de desastres*. Colima, Universidad de Colima.

Macías, J. M. (1999) *Riesgo voclánico y evacuación como respuesta en el Volcán de Fuego de Colima*. Mexico, CIESAS, Universidad de Colima.

Macías, J. M. & Calderón, G. (coords.) (1994) *Desastre en Guadalajara: notas preliminares y testimonios*. Mexico, CIESAS.

Maskrey, A. (1993) Presentación. In: Maskrey, A. (comp.). *Los desastres no son naturales*, Bogota, LA RED-Tercer Mundo Editores, pp. XI–XIII.

Mauch, C. & Pfister, C. (eds.). (2009) *Natural Disasters, Cultural Responses: Case Studies Toward a Global Environmental History*. Lanham, MD, Lexington Books.

Mercer, J. (2010) Disaster Risk Reduction or Climate Change Adaptation: Are We Reinventing the Wheel? *Journal of International Development*. 22 (2), pp. 247–264.

Molina, A. (2014) *Guerra, tifo y cerco sanitario en la ciudad de México. 1911–1917*. Mexico, CIESAS.

Molina, A. (1996) *Por voluntad divina, epidemias y otras calamidades en la Ciudad de México, 1700–1762*. Mexico, CIESAS.

Molina, A., Márquez, L. & Pardo, C. P. (2013) *El miedo a morir. Endemias, epidemias y pandemias en México: análisis de larga duración*. Mexico, CIESAS, Instituto de Investigaciones Dr. José María Luis Mora, Benemérita Universidad Autónoma de Puebla.

Oliver-Smith, A. (2017) Adaptation, Vulnerability and Resilience: Contested Concepts in the Anthropology of Climate Change. In: Kopnina, H. & Shoreman-Ouimet, E. (eds.). *Routledge Handbook of Environmental Anthropology*. London, Routledge, pp. 206–218.

Oliver-Smith, A. (2015) Hazards and Disaster Research in Contemporary Anthropology. In: Wright, J. D. (ed.). *International Encyclopedia of the Social and Behavioral Sciences*. Amsterdam, Elsevier, 2nd edition, pp. 546–553.

Oliver-Smith, A. (2013) Disaster Risk Reduction and Climate Change Adaptation: The View from Applied Anthropology. 2013 Malinowski Award Lecture. *Human Organization*. 72 (4), pp. 275–282.

Oliver-Smith, A. & Hoffman, S. M. (2002) Introduction: Why Anthropologists Should Study Disasters. In: Hoffman, S. M. & Oliver-Smith, A. (eds.). *Catastrophe & Culture: The Anthropology of Disaster*. Santa Fe, Nuevo México, School of American Research, pp. 3–22.

Oliver-Smith, A. (2002) Theorizing Disasters: Nature, Power, and Culture. In: Hoffman, S. M. & Oliver-Smith, A. (eds.). *Catastrophe & Culture: The Anthropology of Disaster*. Santa Fe, Nuevo México, School of American Research, pp. 23–47.

Oliver-Smith, A. & Hoffman, S. M. (eds.). (1999) *The Angry Earth: Disaster in Anthropological Perspective*. New York and London, Routledge.

Padilla, C. & Macías, J. M. (coords.). (1993) *Analizando el desastre de Guadalajara*. Mexico, CIESAS.

Padilla, R. (2018) Prácticas históricas de alertamiento y protección ante huracanes en Baja California Sur. In: Altez, R. & Campos Goenaga, I. (eds.). *Antropologia, historia y vulnerabilidad. Miradas diversas*. Zamora, Mexico, El Colegio de Michoacán, pp. 141–166.

Padilla, R. (2014) *Estrategias adaptativas ante riesgos por huracanes en Cuyutlán, Colima y San José del Cabo, Baja California Sur en el siglo XX*. PhD Thesis in Anthropology. Mexico City, CIESAS.

Peniche, P. (2010) *Tiempos aciagos. Las calamidades y el cambio social del siglo XVIII entre los mayas de Yucatán*. Mexico, CIESAS.

Revista Mexicana de Sociología. (2018) *Sismos: 1985 y 2017*. Special Issue.

Reyes, L. (2007) *Los zoques del volcán*. Mexico, Comisión para el Desarrollo de los Pueblos Indígenas.

Robinson, S., Hernandez Franco, Y., Mata Castrejón, R. & Bernard, H. (1986) It Shook Again: The Mexico City Earthquake of 1985. In: Oliver-Smith, A. (ed.). *Natural Disasters and Cultural Responses*. Williambsburg, Studies in Third World Societies, pp. 81–122.

Rodríguez, J. M. (2007) *La construcción social del riesgo de desastre: ENSO (El Niño/Southern Oscillation) en la cuenca del río Tijuana*. PhD Thesis in Anthropology. Guadalajara, CIESAS.

Rodríguez, M. N. (2018) *Convivir con la amenaza: vulnerabilidad y riesgo frente a los huracanes en la ciudad de Chetumal, Quintana Roo*. Master's Thesis in Social Anthropology. Mexico City, CIESAS.

Rojas, T., Pérez, J. M. & García-Acosta, V. (1987) *Y volvió a temblar . . . Cronología de los sismos en México (de 1 pedernal a 1821)*. Mexico, CIESAS.

Soares, D., Millán, G. & Gutiérrez, I. (coords.). (2014) *Reflexiones y expresiones de la vulnerabilidad en el sureste de México*. Mexico, Instituto Mexicano de Tecnología del Agua.

Steward, J. H. (1955) *The Theory of Culture Change: The Methodology of Multilinear Evolution*. Urbana, University of Illinois Press.

Sutton, M. Q. & Anderson, E. N. (2018) Cultural Ecology. In: Callan, H. (ed.). *The International Encyclopedia of Anthropology*. New York, John Wiley & Sons, Ltd. DOI:10.1002/9781118924396.wbiea1452.

Toledo, V. M. & Barrera-Bassols, N. (2008) *La memoria biocultural. La importancia ecológica de las sabidurías tradicionales*. Barcelona, Icaria editorial.

Torry, W. I. (1979) Anthropological Studies in Hazardous Environments: Past Trends and New Horizons. *Current Anthropology*. 20 (3), pp. 517–529.

UNISDR. (2013) *From Shared Risk to Shared Value: The Business Case for Disaster Risk Reduction*. Global Assessment Report on Disaster Risk Reduction. Geneva, United Nations Office for Disaster Risk Reduction.

Valencio, N. & Siena, M. (orgs.). (2014) *Sociologia dos desastres. Construção, interfaces e perspectivas*, vol. IV. São Carlos, RiMa Editora.

Veland, S., Howitt, R., Dominey-Howes, D., Thomalla, F. & Houston, D. (2013) Procedural Vulnerability: Understanding Environmental Change in a Remote Indigenous Community. *Global Environmental Change*. 23, pp. 314–326.

Vera, G. (coord.). (2009) Devastación y éxodo. Memoria de seminarios sobre reubicaciones en México. Mexico, CIESAS.

Vera, G. (2005a) Vulnerabilidad social y expresiones del desastre en el distrito de Pochutla, Oaxaca. In: García-Acosta, V. (ed.). *La construcción social de riesgos y el huracán Paulina*. Mexico, CIESAS. pp. 35–151.

Vera, G. (2005b) Ancianos, tiemperos y otras figuras de autoridad en dos comunidades del volcán Popocatépetl. La otra visión del riesgo volcánico. In: Macías, J. M. (coord.) *La disputa por el riesgo en el volcán Popocatépetl*. Mexico, CIESAS, pp. 99–116.

Villegas, C. (2005) Recuperando el paraíso perdido: el proceso de reconstrucción en la ciudad de Acapulco. In: García-Acosta, V. (ed.). *La construcción social de riesgos y el huracán Paulina*. Mexico, CIESAS, pp. 153–256.

7 Is there an Anthropology of Risks and Disasters in Peru?

Fernando Bravo Alarcón

Introduction

This chapter takes as a starting point the following paradox: on one hand, there is scientific consensus on the conditions of vulnerability prevailing in Peruvian territory and affecting its population, highly exposed to natural hazards and disasters; on the other hand, local anthropology has not developed an academic interest to assess the impacts made on the symbolic, cultural, and material dimensions of Peruvian society.

This contradiction seems to develop more complicated features when the topics related to risks and disasters have been monopolized by the engineering disciplines both in Peru and other Latin American countries, due to the prevalence of what has been called the physicalist approach or the applied sciences approach (Marino, 2015: p. 32; Gellert, 2012: p. 14; López, 1999; Torrico et al., 2008).

Even though disasters have been a subject of study within a social science perspective for more than nine decades, starting with the publication by Samuel Prince on the social impact of a terrible ammunition depot explosion at Halifax Port (Canada) in 1917 (Oliver-Smith & Hoffman, 1999: p. 18), subsequent research has not always included sociocultural patterns, the historical context of the society under analysis or its vulnerability conditions. Just in the 1970s in Latin America, after a series of seismic events with broad social and physical impacts in this region, a few anthropologists and geographers started to wonder what made disasters much more serious in Latin America than in more developed economies, and they suggest that most likely the impact of natural disasters fell harder on underprivileged social sectors (Oliver-Smith, 2016: p. 109). Thus, little by little, the need to delimitate social processes and structures within a development approach, a relation expressed particularly by disasters, and which would help pave the way to include social sciences in the study of disasters, slowly appeared in the region (Lavell, 1993: p. 80).

But the interest of Latin American anthropology in such events/processes was based in the international evolvement of this subdiscipline in the United States of America and Europe where American and European anthropologists interacted with Latin American anthropologists. Susann Baez Ullberg (2017: p. 3) holds that while risks and disasters constitute a still incipient field of research in the sub-continent, there are a few exceptional cases that appeared rather early. Such is the case of Cuban Fernando Ortiz, who in 1947 presented in a publication an analysis of the mythology and symbology created by some communities around the occur-rence of hurricanes in the Caribbean (Ortiz, 1947). Baez Ullberg also brings up the work and career of already-cited anthropologist Anthony Oliver-Smith,[1] who started in the field with his study on the impact and reconstruction of a city after the 1970 earthquake and landslide in Ancash, Peru.

In 1987, after a series of calamities damaged many towns in different continents, and a growing number of scholars reviewed the paradigms used to understand disasters, the United Nations,[2] declared the 1990s as the International Decade for Natural Disaster Reduction.[3] In fact, seismic events such as the 1985 earthquake in Mexico City became crucial contexts where researchers, government officials, and civil society members acknowledged the pertinence of applied anthropology and other social sciences (sociology, geography, urban planning) for the mitigation of disasters and their socioenvironmental effects (Faas & Barrios, 2015: p. 288). Another relevant event mentioned by Baez Ullberg is the establishment of the Network of Social Studies on Disaster Prevention in Latin America (LA RED) in 1992 in Puerto Limón, Costa Rica, where many anthropologists engaged in talks with other social scientists with the aim of strengthening a more comprehensive approach to such critical events. Thus, academics such as Virginia García-Acosta, Andrew Maskrey, Elizabeth Mansilla, Omar Darío Cardona, Allan Lavell, Jesús Manuel Macías, and Gustavo Wilches-Chaux (Faas & Barrios, 2015: p. 288), among others, became part of an intellectual community which soon encouraged the study of disasters from a social rather than engineering point of view.

It is also true that the development and widening of this perspective has not been uniform in the region, neither in scope nor in depth. Even though there are countries such as Argentina, Brazil, Chile, Mexico, and Venezuela where there are certain research teams, the efforts for strengthening and achieving fluid academic exchanges are still under way.

As for the Peruvian anthropology community, even though it was enriched by the contribution of the already-cited authors and initiatives, it lacked sufficient awareness and interest to develop an academic interest in calamities. Therefore, in order to learn about the causes that prevented Peruvian anthropology from devel-oping a study line on risks and disasters, this chapter starts by describing the academic journey made by Peruvian anthropology. Then, it highlights the expo-sure and vulnerability of Peruvian territory to natural and technological calamities such as earthquakes, floods, drought, heavy rainfall, landslides, contaminating spills, among others. After that, it reviews studies on risks and disasters in Peru conducted from the perspective of history, economy, geography, engineering, and anthropology. Next, there is an attempt to answer the question on why there is not a line of research to study disasters as essentially social events/processes in Peruvian anthropology.[4] Lastly, some actions are suggested to foster a risks and disaster approach in Peruvian Anthropology.[5]

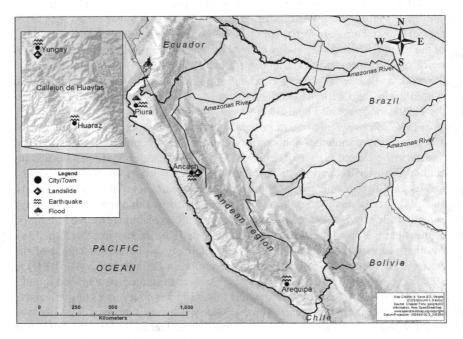

Map 7.1 Map Peru. Case studies and main areas mentioned

Anthropology, risks and territory in Peru

Historical sketch of Peruvian anthropology

Even though chroniclers and travelers during the Spanish Viceroyalty in Peru made observations and descriptions which might be considered as precursors of some kind of anthropology which tried to understand "the other", we may say that this discipline had its academic and institutional beginnings in Peru under the impulse of the intellectual and artistic movement known as the Indigenism. Developed between the 1920s and 1940s, Indigenism made the indigenous world the center of reflection, with manifestations in politics, narrative, visual arts, and photography, and it contributed to the erosion of the excluding paradigm that left the indigenous peoples of the country out of the "imagined national community" (Anderson, 1993). Rodrigo Montoya (2016: p. 19) says, "Indigenism is a chapter of Peruvian political history, and a moment, the first one, in Anthropology".

With the introduction of the professional career of Anthropology at Universidad Nacional de San Antonio Abad del Cusco (San Antonio Abad National University) in Cusco in 1941,[6] and the Institute of Ethnology and Archeology at Universidad Nacional Mayor de San Marcos (San Marcos National University) in 1946, sponsored by ethnologist Luis E. Valcárcel, anthropology started its progressive journey towards institutionalization. Although Peruvian anthropology emerged within public universities, it did not have strong ties with the state until the 1960s, differently from, for instance, Mexican anthropology, which was deeply connected to the Mexican state. In the Peruvian case, it was rather further to the activities of American philanthropic foundations such as the Rockefeller Foundation, the Smithsonian Institution or the Social Science Research Council, and European academic institutions such as the French Institute of Andean Studies (IFEA), created in 1948 (Degregori & Sandoval, 2007: p. 309). Examples of this were the applied anthropology projects in Vicos (Ancash region), started in 1951, and in the Chancay Valley (Lima section near the Andes), in 1964, both projects supported by Cornell University.

Until the 1960s, the primary topics of Peruvian anthropology were studies of indigenous communities, folklore, and applied anthropology projects (Degregori & Sandoval, 2007: p. 310) under the influence of culturalism and still of Indigenism, but also of Robert Redfield and the notions of community development stemming from American anthropology, all of which made cultural change the pillar that underlay across the research of that time (Urrutia, 1992: p. 7).

However, new lines of study began to appear, such as urban Anthropology and the studies on migrants, shanty towns, and provincial clubs; ethnohistory, which reformulates the knowledge of 16th century within the context of the Spanish arrival in Peru; and Amazonian studies dealing with academically little-known spaces and territories. Under the influence of the theory of dependence, which highlighted subject matters neglected by the discipline such as domination, conflict, and power, restricted descriptions formerly considered as primary methodology in fieldwork were left aside to search for interpretations with a national scope (Degregori, 2005: pp. 43–44).

Another event that involved many anthropologists was the agrarian reform (1969–1975) as they joined the government to follow this process and enforce it in the Peruvian countryside (Mayer, 2009). Thus, under the influence of structuralism, and certain Marxist economic viewpoint, studies on peasant communities became relevant in most monographs, displacing the *indio* (indigenous inhabitant) as a subject of analysis (Urrutia, 1992: p. 11).

Until then, Peruvian anthropology was characterized by a few striking contradictions, due to the influence from Europe and North America, apart from the proto-anthropology produced by chroniclers, missionaries, and colonial officials. Similarly, this kind of anthropology became the "spotlight of the ideological debate around our nationality (the question of the indigenous people, miscegenation, and the national identity)" (Falla, 1986: p. 241) but at the same time, it remained a marginal discipline because it was not able to create spaces within the Peruvian government.

In the 1980s, with the outburst of subversive violence, there were some voices criticizing anthropologists for not being capable of properly foreseeing the conditions that made it possible the emergence of the lethal insurgent group of Shining Path in the Andes (Starn, 1992: p. 17; Renique, 1992). At the same time, there were a growing number of studies on ethnicity, gender, and social movements within a "city anthropology" focused on migrations from the countryside to the cities: the search for "the other" shifted from the Andean or Amazon indigenous communities to the urban environment. A thematic diversification takes place along with a shift from anonymous and impersonal structures to actors with a face and individuality, which some described using the metaphor of "the Chorus Rebellion". As Carlos I. Degregori and Pablo Sandoval (2007: p. 319) point out, there is a wide theme shift towards new issues such as identities and mindsets, popular urban culture, religiosity, violence, youth organizations and identities, and consumption, among others.

Meanwhile, nongovernmental organizations became the natural fields of anthropological professional practice in projects of social and cultural promotion under the new approach to multiculturalism and interculturalism. Thus, subject matters such as ecology, sustainable development, and education anthropology were engaged with the professional and academic interests of many anthropologists. Then, with the arrival of the market economy in the country during the 1990s, but above all with the mining boom of early 21st century, many anthropologists turned their professional activities to the private businesses, either as experts in community relations within extractive industries or as managers or consultants in cultural industries. In the academic field, the new themes were visual anthropology, Amazon anthropology, tourism anthropology, territorial anthropology, etc.

As it may be seen, the main topics of Peruvian anthropology have evolved driven by a series of different factors, some of them out of the control or decision of their agents, whether academicians or practitioners. Moreover, this line of work has so far stayed clear of certain topics such as risks and disasters and many others.[7]

It has always been possible for the discipline to develop more extensively, become more influential, cover a wider thematic range, or be more productive. That is why the reason for an absence of a line of studies concerning risks and disasters should not be approached under a blaming or disapproving light, because, as it has already been said, there are other anthropological matters and approaches that have not been developed in the country either, and which might also be deemed as relevant in an academic and practical sense.

Therefore, the absence of the matter of risks and disasters in Peruvian anthropology does not discredit or belittle it. Local anthropological community has its strengths, standard topics, theme networks, and more or less active advocates, as well as weaknesses and intellectual blindness like any other academic community born inside the so-called peripheral anthropologies (Cardoso de Oliveira, 1997).

Peru, a doubly exposed country

The situation of Peru as a country historically exposed to natural hazards has long been documented. For past events, there are historical descriptions (Huertas, 2009; Seiner, 2002, 2011; MINAM, 2016) depicting the incidence and consequences of calamities which afflicted several populations during colonial times. More recently, the tragic events of 31 May 1970 (earthquake and subsequent landslide in Callejón de Huaylas, with more than 67,000 casualties), the fateful 1982–1983 El Niño Southern Oscillation and its damaging 1997–1998 repetition, the August 2007 earthquake in the southern coast, and the more recent March 2017 climate change event (the so-called Coastal El Niño), which caused great havoc in the central and northern coast of Peru, still linger in the country's collective memory. But, on the other hand, there is another kind of hazard, where critical events may not be so evident: climate change. In fact, there are several publications[8] that present specific accounts and experiences of Andean communities where there is a clear glacial retreat, a decrease of water sources, frequent drought, among others, which is indicative that climate disturbance is already taking place in the Peruvian Andes.

"Traditional" hazards and calamities

As several sources and the own experience of its inhabitants may attest, Peruvian territory is subject to certain hazards and calamities that we may call "traditional" or "recurrent", as they take place with some regularity in many of its regions. The southern part of the country, for instance, is known for its vulnerability to seismic events, a risk that is increased by the presence of a chain of active or semi-active volcanoes that pose a risk on populations living on neighboring areas. Another case is that of the northern and central coast, where river overflows and landslides are frequent in the summer months (January, February, and March), being more intense and devastating during the El Niño climatic disturbance.

Official documentation shows that due to its geographic location and land characteristics, Peru is exposed to a rather frequent occurrence of natural hazards such as earthquakes of various magnitudes, landslides, *huaycos*,[9] mudslides, alluvions, floods, droughts, frosts, hailstorms, snowfalls, cold waves,[10] volcanic eruptions, and El Niño oceanic oscillation with unusual levels of rainfall in the north of the country (PNUD/PMA, 2010: p. 156).

The Presidencia del Consejo de Ministros (Presidency of the Ministers' Council), part of the Executive Branch, holds that certain geophysical and climate factors of our country increase the risks of exposition to disasters. One of these factors is the location of Peru in the Pacific Ring of Fire, which explains seisms and volcanic activities, and another is the Andean Mountain Range, which together with the Pacific Anticyclone, causes climate phenomena, all of which makes it difficult to manage disaster risks (PCM, 2014: p. 8).

To this context, we may add prevailing conditions of population vulnerability related to poverty, socioeconomic and regional inequities, a deficient culture of prevention (Bravo, 2019), the unplanned occupation of unsuitable territory, and the inadequacies of physical infrastructure. Such vulnerabilities are shown, for instance, in the pattern of land use, mostly without planning and control (illegal land occupation), making infrastructure and basic services costly and inadequate. We can add precarious self-construction methods without technical knowledge, and the use of unfit materials, all of which increases its risk (PCM, 2014: p. 19).

Another source of risk comes from an ever-growing series of economic and productive activities, infrastructure and technological equipment, not always authorized, which are happening due to the national economic development. In addition to the development of extractive industries (mining, gas, oil), there are other critical events and contingencies that also reflect the way in which Peruvian society keeps reproducing conditions of risk, causing, for instance, spills of dangerous substances or oil into rivers and lakes, or a risky buildup of mining wastes located near human settlements (known as environmental liabilities) (Bravo, 2015), or accidents in illegal mining activities, etc.

Risks resulting from global environmental change

Climate change will trigger some already-known natural events, which might become more frequent and intense, and cause impact on wider spaces than those historically registered. The rainfall patterns, for instance, might become more intense in areas known by their relative aridness, where floods and overflow of water bodies would happen; or they could show withdrawal in areas where previously there was heavy rainfall, causing drought or desertification. The climate and ecosystem diversity of Peru highly exposes the country to disasters as shown by the presence of seven out of nine characteristics of danger set up by the United Nations Framework Convention on Climate Change: low coastal zones; arid and semiarid zones; zones exposed to floods, droughts, and desertification; fragile mountain ecosystems; zones prone to disasters; zones with high urban air contamination; and economies highly dependent on the income generated by the production and use of fossil fuels (MINAM, 2015: p. 20).

To sum up, Peru is a highly vulnerable country due to structural factors such as poverty and inequity, which add to the current vulnerability of the important ecosystems of the Amazon and the Andes, all of which worsens its risk (MINAM, 2015: p. 20).

Academic approach to risks and disasters in Peru

What has been said on risks and disasters in Peru from an academic point of view? From the perspective of social sciences and humanities there is an academic production dealing with a range of natural events and disasters such as earthquakes which, due to their magnitude and impact, drew the attention of chroniclers and historians; or El Niño climate disturbance, which has been the subject of public attention and concern, especially after its devastating occurrence in 1982–1983.

It is perhaps from history that we have more information. For example, the books from historian Lorenzo Huertas (1987) who managed to obtain one of the first documents of 1578 describing the impact of heavy rainfall caused by El Niño, which devastated cities and the countryside on the northern coast of Peruvian viceroyalty (Trujillo and Saña). This study describes the reaction of the indigenous population not only to the loss of their crops, but also to the pressure by the Spanish *encomenderos* for them to pay taxes, and, above all, it describes their need to migrate.

Likewise, Susana Aldana (1996) examined three cases of calamities that occurred in colonial times and caused economic and productive rearrangements, disruptions in the daily lives of the affected people, as well as the adoption of new strategies by some groups who saw they could take advantage of these situations and obtain benefits and gains: the 1687 and 1746 earthquakes and the heavy rainfall that devastated the coastal city of Piura.

A large scope collective work, where several cases of disasters taken place in Latin American countries as well as in pre-Hispanic, colonial and 21st century Peru, is the three-volume *History and Disasters in Latin America* (García Acosta, 1996, 1997, 2008).[11] In the first volume, by using the work of the famous indigenous chronicler Felipe Guamán Poma de Ayala, Lupe Camino identifies the concept of disaster used in the Andean world in the context of colonization, showing that even though the main idea is to consider a disaster a "divine punishment", this does not exclude human participation in its causes to try to reduce its occurrence. In the second volume, Linda Manzanilla identifies a series of material evidence showing the efforts made by pre-Columbian societies located in the Andes to counter the impacts of El Niño. Next, Michael Moseley also focuses on the Andes, where he highlights the double exposure to two powerful sources of physical transformation: the dynamics of tectonic plates and the sun, which manifest itself through climate, radiation, and ocean-atmosphere interplay. Thus, by using the concept of convergent catastrophes, he reviews the way ancient Andean cultures thrive or disappear as a consequence of such disasters.

Besides, Charles Walker (2004) describes the commotion caused in colonial Lima by some rumors circulated in 1756 about the impending city destruction by the fall of fireballs. In a context where magical-religious explanations prevailed, the announcement of calamities and the people's reactions revealed the mentalities, collective fears, and predominant ideas about catastrophes. In his book, Walker (2012) makes the earthquake that hit Lima in 1746 his object of analysis for the social tensions, political alliances, and the circumstances of this city at colonial times. This work is part of contemporary historiography which focuses on incidents or on the study of events as key to describe and understand, on one hand, the disorder itself (in this case the referred seism), and, on the other hand, the subsequent balance, where the author examines the social, political, religious, and mental collective imagination of the different actors reacting to the phenomenon. In other words, it does not see disasters as isolated events but as the result of processes, where baroque piety, reconstruction plans and social control, rebellious feelings of indigenous populations against Spanish rule, and the Bourbon reforms appear very clearly for the benefit of the history student.

Similarly, Yony Amanqui (2008) dealt with the identification of several accounts given about the causes of earthquakes in the city of Arequipa during colonial times, which went from natural explanations (hot exhalations from inside the Earth, accumulated electricity) to those coming from religion and the supernatural (divine punishment), the latter being the more widespread. It should be noted that by assessing crucial events such as tsunamis and earthquakes, this work was able to capture part of the mindset of the time, and the prevailing tensions of the moment.

Another historian who has devoted his efforts to the study of seismic activity and climatic events is Lizardo Seiner (2001), who examined El Niño risks and effects, and concluded that it is necessary to get to know its historical recurrence as a part of a disaster mitigation strategy. In another longer paper, Seiner (2002) reviewed the social impact of critical events resulting from earthquakes, El Niño, environmental pollution, and deforestation between the 16th and early 20th centuries, but also the role of meteorologists and engineers in their respective fields of intervention. In Seiner (2011), the author presents a list of seisms recorded between the 18th and 19th centuries in Peruvian territory through exhaustive research of historical primary sources, reporting that in such period there happened to be 3.304 seismic events of different magnitudes.

Carlos Carcelén (2011) studied three different natural events, all of them in colonial times, namely the 1791–1794 El Niño, the 1746 earthquake in Lima, and the epidemic that ravaged the southern Andes of Peru between 1714 and 1720. Just like Amanqui (2008), Carcelén also finds the religious explanations for these events to be most recurring, while the scientific explanations were a minority.

There are certainly more publications which from the field of history have covered natural disasters and their impact on different cities of the country at different times. A compilation by Ruth Rosas (2001), for example, gathered a group of experienced historians around the timeline of the El Niño phenomenon on the northern coast of Peru to discuss legends and myths originated by this weather event, the calamities inflicted on the cities of the region, as well as the existence of document sources of colonial times.

Pablo Emilio Pérez-Mallaína (2001) produce one of the very first historical pieces of research around a natural event as an observation tool for a social crisis.[12] The quake that happened in Lima in 1746[13] created social and political tensions among the relevant colonial actors around the reconstruction and recovery of the city. Viceroy Manso de Velasco, the clergy, and the families of dignitaries engaged in disputes about the best way to reconstruct infrastructure and recover the urban dynamics using recurrent supernatural and eschatological explanations that were very common at that time. By using the criterion developed by Oliver-Smith on the fact that disasters and reconstruction processes can make power relationships in a society visible, Pérez-Mallaína manages to discriminate among some consequences attributed to the quake itself and others, ruling out that business losses were so large for the city. These and other findings were summarized in a later article (Pérez-Mallaína, 2005, 2008). For his part, Víctor Álvarez (2019) reviews the case of the great earthquake of 1970 that occurred in the Ancash region, whose impact led to an enormous deployment of international solidarity, becoming a catalyst that contributed to the global institutionalization of humanitarian aid in cases of disasters.

Paula Rivasplata (2015) recounts the technological strategies used by Lima authorities during 16th, 17th, and 18th centuries to hold back the water from the fast-flowing Rimac River, and protect the city at the rising seasons. The construction of retaining walls (*tajamares*) alongside the riverbanks required large resources and became a public need to control the river and avoid its destructive impact. In spite of this experience, modern Lima has not been able to solve this problem since the river waters still represent a hazard for the large, low-income population living at its banks.

Concerning geography, there are also publications dealing with cases and subject matters related to risks and disasters. Andreas Haller (2010) examines the process of urbanization and risk perception in the case of the city of Yungay, destroyed by an avalanche of ice, debris, and rocks formed when a side of the Huascaran Mountain collapsed in 1970. Hildegardo Córdova-Aguilar (2017) has studied the vulnerability and risks of human settlements in the outskirts of Lima, which may worsen in case of disturbance of the rainfall patterns due to the climate change. Those areas, in addition to difficulties for water supply, are poorly prepared to stand in case of heavy rainfall. Additionally, Edwin Campos (2017) applied a survey to some Lima school students to find out if there is some kind of statistical relation between the disaster risk management plan and the students' environmental culture, finding that the relationship is not very significant, so there must be other factors in play.

From the economic field, authors like Joanna Kámiche and Aída Pacheco (2010) stressed that there is no research in the country estimating the economic costs of natural disasters for Peruvian homes, and they assessed the loss of assets and/or income caused by those events. Máximo Vega-Centeno (2011), for his part, with the aim of reducing the economic impact of earthquakes on the low-income population, suggests setting up a decision conceptual framework for prevention measures, since the lack of this tool is a huge disadvantage for human development. Many years before, Vega-Centeno and Maria Antonia Remenyi (1984) proposed a methodological scheme to make an economic analysis of earthquakes. Alex Contreras and coauthors (Contreras et al., 2015) present economic estimates of the impacts of El Niño and seismic activity on Peruvian economy, productive sectors, and families.

Finally, the sciences of engineering, architecture, and climate have gained important media exposure, and their approaches are predominant concerning the study of disasters in Peru. The media exposure of experts such as the engineers Julio Kuroiwa (with many articles in the journal *El Ingeniero Civil*),[14] and Hernando Tavera, or the expert in atmospheric sciences Ken Takahashi, shows how the paradigm derived from applied sciences, also known as physicalist, has seeped into the common sense becoming the primary approach par excellence to analyze the degree of exposure of Peruvian society to disastrous events. In the governmental plans against disasters, in the media, and even in the social imaginary, the matters related to calamities and disasters are associated to engineers because of the infrastructure works, calculation and estimates, and material resistance involved, and the social pressure in order to get more accurate and precise predictions.[15]

To say it in other words, at different levels of government management and public opinion, there is an approach that points at the natural phenomena as the only cause of disasters, all of which tends to legitimize the predominance of basic and applied sciences in understanding the problem (Torrico et al., 2008: p. 23).[16] The predominance of this view may explain in part why social disciplines in Peru, anthropology among them, are not taken into consideration at the moment of designing responses, strategies, and plans to face natural events and disasters. In the view of the advocates of physicalism and applied sciences, the anthropological contribution would not be very specific or useful, and they count on the advantage that government agencies dealing with disasters also prioritize engineering and reconstruction aspects.[17]

And anthropology?

On the side of academic anthropology, some practitioners have had episodic contacts with risks and disasters, while others, with longer tenure, have moved between the academic and the professional fields. In the former case, as a result of his involvement in the emergency squads to help the victims of 1970 Yungay and Huaraz earthquake, anthropologist Rodrigo Montoya specifically wrote his reflections and observations around his personal experience. It is particularly interesting to see how such calamity made more visible ethnic and social gaps in those places, when middle-class families (teachers, small businessmen, employees, small property owners) thought they were the most affected victims. In correlation with that perception, those sectors received more aid than "the indians". Montoya points out the discomfort felt at the time of "having to queue to receive food together with the Quechua farmers" (Montoya, 1970: p. 8). The disaster was an event that revealed the class tensions and differences.

A different and subsequent case is that of the late Peruvian anthropologist Eduardo Franco Temple, who was an expert in the study of natural events and disaster risk management from a social perspective. From the NGO Intermediate Technology Development Group, and LA RED,[18] Franco not only was professionally engaged in this matter, but also took part in academic and scientific research from a disaster risk perspective, about which he made many presentations and publications (Franco, 1991, 2000, 2006; Franco & Zilbert, 1996). He was also the editor of the journal *Desastres y Sociedad* and a member of the editorial staff of the journal *Tecnología y Sociedad*, where he published reports on El Niño, prevention systems, and civil defense systems.[19]

Another contribution to the study of disasters was the article by Spanish anthropologist Carlos Junquera (2002), who worked the case of El Niño that affected the northern city of Piura, and supported his anthropological viewpoint about disasters, developing focuses, topics, trends, cases, and the theoretical evolvement of this specialty. It should be noted that this article was published in a Peruvian journal of geography, not of anthropology.[20]

A noteworthy case is that of American anthropologist Oliver-Smith,[21] who, in 1966, as a young graduate visited Yungay, a city in the *Callejón de Huaylas*, in Ancash, to do some research after graduating from the Indiana University. He had plans to return there early in 1970 to collect information for his doctoral thesis, but the tragic 31 May seism and the ensuing mudflow forced him to postpone his return until months later. With the complete alteration of his anthropological interest, and following the advice of his supervisor, Oliver-Smith felt the need to make a shift in his research: he decided to document the recovery and reconstruction process of Yungay, the town destroyed by the lethal event (Oliver-Smith, 1979, 2015b), and that work would become a milestone in the development of risk of disaster anthropology in Latin America (Oliver-Smith, 2002: p. 147, 2016: p. 109; Faas & Barrios, 2015: pp. 287–288).[22] His observations and conclusions on this experience were presented in his 1986 book *The Martyred City: Death and Rebirth in the Andes* (Oliver-Smith, 1986). In his own words: "This book was re-edited in 1992 and tells the story of the survival process of Yungay, of how its inhabitants fought against huge forces so that their traditions would not disappear. Unfortunately, it is not available in Spanish" (Oliver-Smith, 2015b).

But Oliver-Smith was not the only researcher interested in the 1970 disaster. Another anthropologist who worked in the area of the catastrophe was Barbara Bode, an American who initially had intended to study changes in local religious beliefs, but after the calamity she decided to shift to other interests. Thus, she set up her headquarters in the city of Huaraz, the capital of the Ancash region, 56 kilometers south of Yungay, and proceeded to prepare a detailed ethnography of the disaster by describing the emotional shock the quake inflicted on the city, as well the process of reconstruction, and the survivors' expectations and aspirations, highlighting, for instance, the tensions with the central government, then led by a "nationalist and revolutionary" government (Bode, 2015).[23]

Going back to Oliver-Smith, after his experience in Yungay, the researcher opted for the study of the relationship between the social conditions of vulnerability, underdevelopment, and the occurrence of critical events, practically beginning a whole new specialty in anthropological studies, which started to gain validity and legitimacy with the succession of research papers and publications where he showed the pertinence for anthropology to understand and explain the way communities deal with risks and disasters. Two relevant articles on the Peruvian case were *Post disaster consensus and conflict in a traditional society: the 1970 avalanche of Yungay, Peru*, published in 1979, and *Peru's five hundred year earthquake: vulnerability in historical context*, in 1994, besides other publications where he deepened and broadened his perspective on post-disaster aid and reconstruction and the analysis of resettlement projects, among other things (Oliver-Smith, 2002, 2015a, 2016, 2017).[24] This way, by retrieving the interest of other anthropologists who studied past experiences of disaster, such as Raymond Firth, James Spillius, or David Schneider, Oliver-Smith took up this line of study, and made it relevant in the academic and applied fields.

Other foreign researchers who have studied events related to other kind of disasters, such as the El Niño oceanic oscillation, are anthropologist Anne-Marie Hocquenghem and paleo-climatologist Luc Ortlieb, who clearly proposed an alternative chronology for El Niño (Quinn et al., 1987), because the other predictions were not verifiable, and in other cases there were doubts on the reliability of their transcriptions (Hocquenghem & Ortlieb, 1992: pp. 198–199).

Finally, we have not been able to find any written publication from an anthropological perspective dealing with disasters caused by technological agents making impacts on populations and ecosystems, such as the case of oil spills, or waste materials leaks from mining operations.

As the bibliographic review shows, Peruvian anthropology has rather had a limited contact with the risk of disaster approach, mainly from a few individual contributions by primarily foreign anthropologists. Most of the contributions have come from historians and geographers, as well as engineers and economists,[25] which confirms what Oliver-Smith maintained (2016: p. 105) when he said that "most research on disasters in Peru has not been conducted by anthropologists".[26]

Why has risk of disaster anthropology not been able to establish itself in Peru, even though there are qualified and competent professionals?

In what follows, we will try to identify a number of possible reasons to under-stand the said situation in a sort of provisional hypothesis. One first explanation is based on the already-cited characteristic of Peruvian society and territory, highly exposed to risks, hazards, and disasters, which is expressed in the occurrence of many geophysical or climatic events with huge impacts on the physical infrastruc-ture and the social dynamics.[27] Thus, the prevailing group of ideas, assumptions, and concepts on those situations, which consider risk as a result of the interplay of extreme geophysical agents, sicknesses, technological failures – that is, the physicalist paradigm – is still predominant in the country whenever the risk of disasters is discussed. So government authorities, experts, specialized institu-tions, and the media usually prioritize technical, reconstruction, reparation, and preventive aspects when faced with the possibility of disaster. Cultural aspects, the adaptation of the population to their physical milieu, the role of social beliefs and moral values, the relationship between social conditions, and vulnerability, in general, are set aside in the background, because the urgent needs come first (material support to victims, health care, resources for reconstruction, recovery of infrastructure, etc.).

As a logical consequence, the hegemony[28] of the physicalist or risk paradigm,[29] brings about the predominance of exact disciplines concerning risks and disasters: engineers, seismologists, meteorologists, climatologists, urbanists, and architects become the experts who receive most of the attention by public managers, the media, and society in general. These experts take advantage of some characteris-tics of their sciences, such as accuracy (the quake had a magnitude of X degrees), some prediction capability (as for El Niño, for instance), or damage estimations (such event might cause X number of casualties), information that is very valuable for decision-making. Thus, these sciences being hegemonic in the field of risks and disasters, and having technical expertise priority over social knowledge, the incentives are lost for social disciplines to introduce themselves in these areas with ease. That explains the reduced presence of sociologists and anthropologists in the local community of disaster specialists, both in the public and the private sectors. In a personal talk with anthropologist Lucy Harman, Emergency and Risk Manager for NGO Care Peru, she pointed out the surprise she noticed in her con-versational partners when they found out she was an anthropologist, an indication of a generalized perception that such professionals do not have much to do or contribute on the subject of disasters.

Another explanation to take into account is the influence of "the major issues" in Peruvian anthropology (the Indians, the farmer community, the rural world, migrations, social change, etc.), which monopolized the interest of new profes-sionals setting other approaches aside – perhaps "less anthropological" prima facie – to more secondary positions. In a personal interview with anthropolo-gist Carlos Aramburú, he noted that, for instance, among his colleagues from the Pontificia Universidad Católica del Perú (Pontifical Catholic University of Peru) an essentialist approach to cultural anthropology was prevalent, which made it difficult to see "circumstantial" issues. As a kind of hypothesis, we can say that "the great topics" of Peruvian anthropology overshadowed the interest in less attractive themes; this helps to understand why disasters were taken for granted as a nonanthropological subject, more related to engineering practice, humanitarian activities, and government policies, rather than an object of academic interest.

When asked about the possible reasons to explain the absence of risk and disaster anthropology in Peru, Oliver-Smith highlighted the weight of certain topics and practices in the course of the profession, as well as the political situation it had to endure. For instance, when he started his studies in Yungay in 1970, Peruvian anthropology was moving away from applied anthropology, and it began to study the ethnography of the Andean world and Andean communities under the approach of cultural anthropology. Besides, in those years, social sciences in general, and anthropology in particular in Peru were more interested in the study of the means of production and the articulation of farming economy with capitalism in a context where the topics of development and underdevelopment were in the spotlight. On the other hand, Oliver-Smith also pointed out the impact of the terrorist action by Shining Path on the anthropological research conducted in the Andes during the 1980s, making it difficult for the field study, and directing the research towards political issues and the role of the farming population in political activism and revolution.

Likewise, Oliver-Smith acknowledges that delays in the publishing of one of his main books, dealing with the already-cited 1970 disaster (Oliver-Smith, 1986), which was never translated into Spanish, as well as the extension of his interest to other topics and fields of study, may have hindered the emergence of this new perspective. The success of LA RED was not either strong enough to boost the topic of disasters in the interests of Peruvian anthropologists, who in the 1990s were more oriented to other subject matters.

Another point that should also be taken into consideration is the absence of charismatic personalities who might have fostered the interest in risks and disasters at the local departments of anthropology. Very often, the connection within an academic community depends on the activism and leadership of an advocate who may call, encourage, and move other colleagues, managing to wake and pass on his/her interest to other researchers and scholars. It usually happens that this agent has resources (contacts, access to finances, proficiency in the matter, etc.) as well as other personal qualities, charisma, and influence among students and fellow anthropologists, all of which may facilitate the formation of networks and think tanks. These groups can share an interest in specific topics, meet regularly to academically discuss their subject of interest, organize events, and do research. There has not been such an articulate personality or motivating small groups at the local departments of anthropology.[30] However, among the several research teams established there, we may find the Research Group on Climate Change and Disaster Risk Management,[31] at the Pontifical Catholic University of Peru, a team that is interdisciplinary in nature, also including anthropologist John Earls as one of its members.[32]

This leads to what happens with the list of courses usually offered at anthropology faculties. A quick review of the curricula and syllabi of courses given at the Pontifical Catholic University of Peru,[33] San Marcos National University,[34] the Universidad Nacional del Centro del Perú (National University of the Center of Peru),[35] and San Antonio Abad National University of Cusco[36] shows the absence of a course developing risks and disasters.[37] Without courses or teachers to impart this specialty, it is clear to see that there is no interest or that there is the false notion that risks and disasters need not an anthropological study.

It is necessary to reiterate that the subject of risks and disasters is not the only matter that Peruvian anthropology has missed. There are also other topics that have not been addressed or received much attention among local anthropologists, namely consumption anthropology, medical anthropology, juridical anthropology, subjectivity anthropology, body and emotions anthropology, sports anthropology, etc. These omissions are not the expression of incompetence or lack of qualification in the academic community in question. Within the context of limitations for the scientific practice in the country (resources, financial support, scholarships, publications, fellowships, etc.), local anthropologists have acted according to the possibilities that reality provides them with. But the Peruvian anthropological community is not so large or connected as it is the case of its counterparts in other countries in the region, let alone the communities in Europe and North America: "We are a small group. . . . We are – and will always be – miles away from academic and professional communities such as those of lawyers, engineers, or accountants" (Diez, 2012: p. 76). Notwithstanding that, and although many subject matters are not addressed within its scope, the quality of local anthropology is unquestionable, and many other topics have had a relatively acceptable treatment; however, it is true that its contribution is needed to enhance the prevailing approaches to disasters in the country, which are, as it has been stated, dominated by more technical professions and the physicalist paradigm of applied sciences.

Finally, and despite this, there are some stimulating facts. For example, at the VIII National Congress of Anthropological Research, held in September 2018 at San Agustín National University of Arequipa, the student Cristian Cadillo Saavedra, from the San Marcos National University, presented his paper entitled "Change and continuity. The process of sociocultural transformation of the town of Yungay after the catastrophe of 1970". Also, in October 2018 the Pan-American Institute of Geography and History, IPGH-Peru organized the seminar "Anthropology of Disasters: towards an agenda for research", where various anthropologists and sociologists presented works on risk, climate change and social responses to disasters. By last, at the 16th Anthropology Student Colloquium held at the Pontifical Catholic University of Peru, in November 2018, there was a panel on societies, devastation, and the local confrontation to disasters, where two students of this university presented the first findings of their research, an event that may signal the emergence of an interest in this field: Maria José Montoya talked on "The State through bureaucracy. The case of 2017 El Niño post-disaster reconstruction" and Deborah Sánchez lectured on "Current use of public space by the inhabitants of flooded areas of the Amazon: the case of 'Munich' human settlement".

How can the risks and disaster approach in Peruvian anthropology be advanced?

Local anthropology should search and foster points of contact with technical and engineering disciplines, which are predominant in the field of risks and disasters, so that the social and anthropological perspective may contribute to the field of vulnerability analysis, post-disaster assistance and reconstruction, and social organization. It is a matter of dialog and articulation with those specialties on an equal footing, showing both the pertinence of the anthropological perspective and its flexibility to interact with academics and experts who respond to another epistemology.

On the other hand, the departments of Anthropology should consider the gradual introduction of courses dealing with topics related to disasters and vulnerability. Additionally, they should organize seminars, lectures, research teams, talks, and other academic events in order to increase the interest and relevance of this approach.

Another needed measure is the determination of potential funding sources for projects, both academic and applied, where the risk and disaster variables may be included. If the academic treatment of these topics is internationally sponsored, the difficulty to access funding would not be harder than other areas of interest.

Although it may seem obvious, it is necessary to establish contacts and exchanges with anthropological communities from other countries of the region with experience in the risk and disasters approach. Topics such as reconstruction, vulnerability analysis, social organization, the experience of population resettlement, impacts of displacement, the attachment to a place, resistance movements, the management of resettlement projects for displaced populations, prevention culture, migrations and disasters, land-use planning, extractive industries and environmental health, and post-disaster recovery are some of the most relevant topics of this approach, which may be enhanced with this interaction.

It should not be ignored that there is a need to encourage teachers to choose the specialization in the risk and disaster approach, be it in courses abroad, fellowships, or postgraduate studies. This not only would help the introduction of this approach in the courses but also the chance to advise on theses prepared within this approach.

It is also imperative recover previous academic production coming from the fields of history, geography, and economy, as well as the accounts of recent critical events and disasters, in order to systematize, study, and turning them into a useful input to support the work of theses, teaching research, or teaching materials for course preparation, etc.

Finally, given the role played by some personalities in the activation of academic interest on a topic – due to their charisma, career, determination, or expertise – and taking into account previous considerations, create the ideal conditions for scholars to assume a driving role and become effective facilitators and advocates for the risks and disaster perspective in Anthropology.

Conclusions

Despite the limitations that historically hinder scientific and academic development in Peru – especially in the field of social sciences – Peruvian anthropology has achieved a relatively important academic and professional institutionalization in the country. The establishment of Anthropology Departments in several public and private universities, research studies on relevant subjects, the publication of studies and magazines, the participation of anthropologists in forums of public and academic debate, and the creation of networks of anthropological studies, among others, are all signs that the local anthropological community shows some vitality.

However, there are still some difficulties related to its small size, the difference between the practice of anthropology in the capital and the practice in the provinces, the different contents studied in public and private universities, the lack of continuity of publications, the weak relationship with other more developed and stronger communities, and the deficient coverage of certain subjects which are studied in more depth in neighboring countries.

One such case is the Anthropology of Risks and Disasters, where while there is some presence of anthropologists who, for academic or professional reasons, have individually engaged in such approach, there is no formal group promoting its study as a subspeciality within the discipline. This is striking in a country showing high levels of social vulnerability to natural hazards and disasters. We are, therefore, in a situation where there are academic skills and competence, as well as some anthropologists with personal experience, but all of this has not created a space for reflection and study leading to thinking of and engagement in such topic.

In spite of this, new generations of anthropologists are showing greater receptiveness towards the subject matter of risks and disasters as seen in recent academic events where, within an anthropological view, natural phenomena are studied when they make an impact on high vulnerability social spheres. However, it is also true that disasters have gotten greater public visibility since its occurrence receives wide media coverage, generating more information and developing a growing citizen pressure for better management, that is to say, there is a favorable context for anthropology to generate hypothesis, study cases, start ethnography, recount experiences, and describe the meanings that societies assign to disasters.

Notes

1 Some of his theoretical works on disasters are Oliver-Smith (2013, 2015a, 2017), Oliver-Smith & Hoffman (1999), and Hoffman & Oliver-Smith (2002).
2 Available from www.un-spider.org/es/riesgos-y-desastres/ONU-y-gesti%C3%B3n-del-riesgo-de-desastres. Accessed 16 November 2018.
3 At that time, the notion that disasters had nature as their primary cause was accepted, therefore, the name "natural disasters" was commonplace in the media, government agencies, and popular speech.

4 Citing anthropologists such as Bruce Kapferer, Lotte Meinert, Anthony Oliver-Smith, and Susanna Hoffman, Baez Ullberg defines disasters as events/processes because, on one side, they are key for the redefinition of social, political, and environmental relationships in those societies where they take place; on the other hand, because they are the outcome of historical development and imply effects both in the present and in the future of a social group (Baez Ullberg, 2017: p. 2).

5 By this, I mean the anthropology practiced by authors from Peruvian academia.

6 From that year until 1999, with the inauguration of the Anthropology Department at Universidad Nacional Federico Villarreal (Federico Villarreal National University), there were ten universities teaching this discipline.

7 Such are the cases of juridical anthropology or medical anthropology. Likewise, there is no development of Environmental Anthropology in Peru, in spite of the situation of its territory as a culturally and biologically diverse space (Bravo, 2013: pp. 74–75).

8 These articles are mentioned in Bravo (2014).

9 According to NGO Predes, with experience in disaster prevention, *huayco* is a Quechua word used in Peru to mean torrential and fast-flowing streams of turbid water running down from the highest part of mountains dragging stones, brushwood, and other sediments along a course of ravines to reach towns and other human settlements. Their direct cause is heavy rainfall taking place during the rainy season. Available from www.predes.org.pe/wp-content/uploads/2017/12/cartilla-huaycos.pdf. Accessed 10 November 2018.

10 The Peruvian National Meteorological and Hydrological Service (Senamhi) defines a cold wave as the weather phenomenon distinguished by the cooling of air coming from Antarctica. Cold winds enter the country by the southern jungle and move to the central and northern jungle, depending on their intensity. They cause a dramatic fall of temperature in the normally hot jungle areas, causing health problems to the populations living there. Available from www.senamhi.gob.pe/?p=heladas-y-friajes-preguntas Accessed 29 October 2018.

11 Virginia García-Acosta, trained in history and anthropology, has several publications where she combines both professions: García-Acosta (2004, 2014, 2018a). For more theoretical aspects of Anthropology of Disasters, see García-Acosta (2005), and for a general review of this subject, see García-Acosta (2018b).

12 The concept of crisis has a deep relationship with disaster, a strong link that is present in everyday language, both colloquial and academic. For a discussion of the concepts of crisis and disaster from an anthropological viewpoint, see Barrios (2017).

13 As it can be easily seen, the numerous references to the 1746 earthquake are a signal that such an event and the social, economic, and political processes it produced have been of particular interest for Peruvian historians and foreign academics who are concerned with the study of disasters: Aldana (1996); Pérez-Mallaína (2001, 2005, 2008); Oliver-Smith (1997); Walker (2012); García-Acosta (2014).

14 Another specialist who contributed to a better understanding of disasters was Alberto Giesecke (Giesecke & Silgado,1981), former director of the Centro Regional de Sismología para América del Sur (Regional Center for Seismology for South America), an engineer who was always keen to study these events by going beyond geology or engineering so that he could highlight their social and economic dimensions (Vega-Centeno, 2011: p. 59). It should be noted that the work by Enrique Silgado (1978) is still today the best chronology of seisms in the history of Peru from 1513 to 1974.

15 More from the field of consultancy firms than from the academia, some recognize the hegemony of a vision centered on disaster itself, that is, on the series of physical, economic, and social damages as a result of a catastrophic event, and which must be met with a swift response. Within this vision, it is a prevalent concept that disasters are essentially natural and dangerous events, hard to prevent and control and therefore the communities would be subjected to the "fury of nature" (Zilbert, 2008: pp. 94–95).

16 This physicalist paradigm of applied sciences takes as premises the following: a) the cause of disasters comes from a physical hazard; b) the objects of main analysis, if not the only one, are the physical aspects of disasters; c) applied sciences such as geology, geophysics, and meteorology provide with a theoretical scientific basis for the study of disasters; d) therefore, the foundation of disaster management, taking physical vulnerability as its base, must focus on measurement and prediction (Torrico et al., 2008: pp. 25–27).

17 There is also literature, not properly academic but coming from government agencies like Instituto Nacional de Defensa Civil (National Institute for Civil Defense), Centro Nacional de Estimación, Prevención y Reducción del Riesgo de Desastres (National Center for Assessment, Prevention and Reduction of Disaster Risks), Programa de Reducción de Vulnerabilidades frente al Evento Recurrente de El Niño (Program for Vulnerability Reduction to El Niño Recurrent Event), or private institutions (NGO Soluciones Prácticas and the Center for Studies and Prevention of Disasters) which from a perspective of public policies and decision-making provide strategies, plans, programs, and experience sharing. All of them provide concepts and categories aimed at support and efficient responses to emergencies and disasters, such as vulnerability analysis, prevention, precaution, reconstruction, resilience, population resettlement, risk assessment, etc. Note that they are categories contributed by risk and disaster anthropology. However, the professional profile prevailing in government institutions is more engineering than social scientific; as for NGOs, there is more professional diversity with a high predominance of sociologists.

18 Note that LA RED had its headquarters in Peru for some years. García-Acosta (2018a) has a summary on this time.

19 Available   from   www.desenredando.org/lared/grupo/EduardoFranco_2001sp.pdf. Accessed 24 September 2018.

20 A review of the journal *Anthopologica* issued by the Catholic University of Peru, a regular publication since 1983, and the journal *Investigaciones Sociales* (Social Research), issued by San Marcos National University, a regular publication since 1995, shows the absence of articles written by anthropologists developing the matter in question, save a pair of comments on books immersed in the subject matter, in the case of *Anthopologica*. Besides, it should be added that, apart from the two cited journals, there are other academic journals, although published irregularly. For instance, see *Revista Peruana de Antropología* (*Peruvian Journal of Anthropology*) published by Centro de Estudios Antropológicos "Luis Eduardo Valcárcel", Universidad Nacional de San Agustín de Arequipa ("Luis Eduardo Valcárcel" Center of Anthropological Studies at San Augustin National University of Arequipa); *Antropología: revista de investigación, análisis y debate* (*Journal of Research, Analysis, and Debate*), at the Universidad Nacional del Altiplano (National University of Altiplano); and *Revista del Museo de Arqueología, Antropología e Historia* (*Journal of the Archeology, Anthropology, and History Museum*) at the Universidad Nacional de Trujillo (National University of Trujillo).

21 It might be said that, to a certain extent, he is an anthropologist with a keen interest in Peruvian culture and history as he published articles and books on disasters and other critical events in Peru where he empirically supported his approach to Risk and Disaster Anthropology.

22 Another American anthropologist who has published with Oliver-Smith and personally experienced the effects of a disaster was Susanna Hoffman. In 1991, due to a raging fire in Oakland, California, Hoffman lost all her personal belongings, and that marked a turning point in her academic interests: "It was very important, it changed everything. I did not even know there was a field of study [risk and disaster anthropology]. It was because of the fire that I began working on this subject" (Díaz, 2017: p. 254). Some of her publications are Hoffman (1998, 1999a, 1999b), as well as those books coedited with Oliver-Smith (Hoffman & Oliver-Smith, 2002; Oliver-Smith & Hoffman, 1999).

23 German anthropologist Martin Sökefeld published an article making comparisons between the 1970 earthquake and the 2010 disaster when a giant landslide hit the people of Hunza in Gilgit-Baltistan (Pakistan). He pointed out how vulnerability to disaster is also worsened by the political conditions prevailing in the affected populations (Sökefeld, 2013).

24 It is noteworthy that, although Oliver-Smith's interest in the disaster approach starts with his experience in Peru, his Peruvian colleagues did not follow suit.

25 This does not necessarily mean that such professions have academic or professional communities specialized in risks and disasters, with the only exception of engineering, where the influence of Julio Kuroiwa is well known.

26 It is necessary to say that the little or non-existing interest of Peruvian academic anthropology in risks and disasters does not mean there are no social intervention projects on disaster prevention where we can find anthropologists in the working teams. For instance, the NGO Pro-DIA association has a project seeking to develop protection mechanisms in the face of disasters for alpaca growers from Puno, with involvement of anthropologists. Another example is the "Proyecto Glaciares +", sponsored by Swiss aid cooperation, aiming at improving the ability to complete adaptation, and the gradual reduction of disaster risks in the face of glacier retreat in Peru. Available from www.proyectoglaciares.pe/proyecto-glaciares/. Accessed 10 December 2018.

27 Taddei (2014), when trying to determine the reasons why disasters did not "exist" for Brazilian anthropology, suggests that this situation changed thanks to three factors: the occurrence of great visibility disasters in the southeastern part of Brazil; the occurrence of international disasters that finally affected Brazil, and about which there was no background; and finally a series of developments in social theory which made disasters a new subject of analysis. It is funny that in Peru the relatively high frequency of disasters has not yet "awaken[ed]" or "raise[d] awareness" in local anthropology.

28 It is certainly not an *absolute* predominance of this perspective over the social approach. In fact, there is awareness that certain social or cultural conditioning must be considered or addressed in disaster prevention. An example of this is the recurrent criticism by social and political communicators on the fact that low-income families carelessly build their homes in high-risk zones, such as riverbanks or at the foot of mountains.

29 Honduran anthropologist Marisa López opposes the "vulnerability paradigm", which states "disasters depend on the social order, daily relationships of society with the environment, and the historical circumstances that characterize the context in which the population lives" (López, 1999: p. 7).

30 Anthropologist Alejandro Diez holds that "Research in any discipline has always required the establishment of work teams or affinity groups who may address related subject matters, providing feedback to our work by criticizing, commenting, but above all, contributing to it" (Diez, 2012: p. 60).

31 This group aims to develop research and train specialists on issues related to climate change from a disaster risk management approach. It is constituted by a group of professionals interested in facing the challenge of climate change in a context of economic growth and expectations of overcoming poverty and social exclusion.

32 Earls' work has tried to articulate anthropology with climate change, the organization of farming communities and the Andean agriculture (Earls, 2009, 2014), Araujo and Earls (2015), Earls and Cervantes (2015).

33 Available from http://facultad.pucp.edu.pe/ciencias-sociales/wp-content/uploads/2018/07/plan-de-estudios-antrologia.pdf. Accessed 16 September 2018.

34 Available from http://csociales.unmsm.edu.pe/images/Reporte_PlanEstudio_antropologia.pdf. Accessed 16 September 2018.

35 Available from www.uncp.edu.pe/sites/uncp.edu/files/pregrado/antropologia/_pdf/plan-antropologia.pdf. Accessed 16 September 2018.

36 Available from http://an.unsaac.edu.pe/home/. Accessed 16 September 2018.
37 It should be said that study programs are flexible and include courses with variable contents that may incorporate the subject matter of risks and disasters, even though that is subject to multiple contingencies (student interest, specialized teaching staff, necessary coordination with other topics, etc.)

References

Aldana, S. (1996) ¿Ocurrencias del tiempo? Fenómenos naturales y sociedad en el Perú colonial. In: García-Acosta, V. (coord.). *Historia y desastres en América Latina*. Bogota, LA RED-CIESAS-Tercer Mundo Editores, vol. I, pp. 123–145.
Álvarez, V. (2019) *El terremoto en el Callejón de Huaylas, Perú y la ayuda humanitaria: un «momento global» durante la Guerra Fría (1970–1973)*. Berlin, Freien Universität Berlin, PhD Thesis.
Amanqui, Y. (2008) Explicaciones sobre las causas de los terremotos dadas en la Arequipa colonial. *Illapa: revista latinoamericana de ciencias sociales*. 3, pp. 253–270.
Anderson, B. (1993) *Comunidades imaginadas. Reflexiones sobre el origen y la difusión del nacionalismo*. Mexico, Fondo de Cultura Económica.
Araujo, H. & Earls, J. (2015) *La gestión del territorio en las comunidades altoandinas y el cambio climático: Investigación-acción en una experiencia de recuperación de terrazas*. II Congreso Internacional de Terrazas: Encuentro de culturas y saberes de terrazas del mundo. Cusco, Centro de Estudios Regionales Andinos Bartolomé de Las Casas, pp. 202–225.
Baez Ullberg, S. (2017) La contribución de la antropología al estudio de crisis y desastres en América Latina. *Iberoamerican–Nordic Journal of Latin American and Caribbean Studies*. 46 (1), pp. 1–5.
Barrios, R. (2017) What Does Catastrophe Reveal for Whom? The Anthropology of Crises and Disasters at the Onset of the Anthropocene. *Annual Review of Anthropology*. 46, pp. 151–166.
Bode, B. (2015) *Las campanas del silencio. Destrucción y creación en los Andes*. Lima, Fondo Editorial del Congreso del Perú.
Bravo, F. (2019) Hacia una cultura de la prevención ante desastres. *Punto Edu*. 463, p. 14.
Bravo, F. (2015) *El pacto fáustico de La Oroya. El derecho a la contaminación "beneficiosa"*. Lima, INTE-PUCP.
Bravo, F. (2014) Las investigaciones sociales sobre el cambio climático. Una revisión preliminar. *Revista Argumentos*. 4, pp. 64–74.
Bravo, F. (2013) Los asuntos ambientales en la teoría antropológica. *Revista Peruana de Antropología*. 1 (1), pp. 67–76.
Campos, E. (2017) Plan de gestión de riesgos de desastres y cultura ambiental: un análisis desde el enfoque cuantitativo. *Espacio y Desarrollo*. 29, pp. 135–151.
Carcelén, C. (2011) Desastres en la historia del Perú: climas, terremotos y epidemias en Lima durante el siglo XVIII. *Investigaciones sociales*. 15 (26), pp. 97–113.
Cardoso de Oliveira, R. (1997) *Antropologías periféricas versus antropologías Centrales* [online]. Available from www.memoria.fahce.unlp.edu.ar/trab_eventos/ev.7091/ev.7091.pdf. Accessed March 5th 2019.
Contreras, A., Martínez, M. & Vásquez, K. (2015) Impactos de El Niño en el Perú. *Moneda*. 164, pp. 28–31.
Córdova-Aguilar, H. (2017) Vulnerabilidad de los asentamientos de la periferia de Lima Metropolitana frente al cambio climático. In: Sánchez, R., Hidalgo, R. & Arenas, F.

(comp.). *Reconociendo las geografías de América Latina y el Caribe*. Santiago, Pontificia Universidad Católica de Chile, pp. 209–233.

Degregori, C. (2005) Panorama de la antropología en el Perú: del estudio del otro a la construcción de un nosotros diverso. In: Degregori, C., Sendón, P. & Sandoval, P. (eds.). *No hay país más diverso: compendio de antropología peruana*. Lima, PUCP-UP-IEP.

Degregori, C. & Sandoval, P. (2007) La antropología en el Perú: del estudio del otro a la construcción de un nosotros diverso. *Revista Colombiana de Antropología*. 43, pp. 299–334.

Díaz, G. (2017) Antropología y desastres: Entrevista con Susanna Hoffman. *Cuhso, Cultura-Hombre-Sociedad*. 27 (2), pp. 251–258.

Diez, A. (2012) La enseñanza de la Antropología en el Perú: escuelas, programas y retos para el desarrollo de la especialidad. In: Rivera, E. (ed.). *Libro memoria. VI Congreso Nacional de Investigaciones en Antropología*. Puno, Escuela Profesional de Antropología, Universidad Nacional del Altiplano.

Earls, J. (2014) Compatibilización de conocimientos climáticos: Una aproximación. In: Damonte, G. & Vila, G. (eds.). *Agenda de investigación en temas socioambientales en el Perú*. Lima, CISEPA-PUCP.

Earls, J. (2009) Organización social y tecnológica de la agricultura andina para la adaptación al cambio climático en cuencas hidrográficas. *Tecnología y Sociedad*. 16, pp. 13–31.

Earls, J. & Cervantes, G. (2015) Inka Cosmology in Moray: Astronomy, Agriculture, and Pilgrimage. In: Shimada, I. (ed.). *The Inka Empire: A Multidisciplinary Approach*. Austin, University of Texas Press.

Faas, A. & Barrios, R. (2015) Applied Anthropology of Risk, Hazards, and Disasters. *Human Organization*. 74 (4), pp. 287–295.

Falla, A. (1986) La antropología en el Perú. *Anthropologica*. 4 (4), pp. 241–245.

Franco, E. (2006) "El Niño" en el Perú: viejos y nuevos temas. *Tecnología y sociedad*. 7, pp. 155–175.

Franco, E. (2000) El Niño en el Perú: Hacia una contextualización de las respuestas sociales al Niño 1997/1998. In: Felipe-Morales, C. & Canziani, J. (eds.). *Mesas regionales impacto de "El Niño": Investigaciones arqueológicas en la costa norte*. Lima, Sepia.

Franco, E. (1991) *El fenómeno El Niño en Piura: ciencia, historia y sociedad*. Piura, Centro de Investigación y Promoción del Campesinado.

Franco, E. & Zilbert, L. (1996) El sistema nacional de defensa civil en el Perú o el problema de la definición del campo de los desastres. In: Lavell, A. & Franco, E. (eds.). *Estado, sociedad y gestión de los desastres en América Latina: en busca del paradigma perdido*. Lima, LA RED-ITDG, pp. 309–441.

García-Acosta, V. (2018a) Vulnerabilidad y desastres. Génesis y alcances de una visión alternativa. In: González de la Rocha, M. & Saraví, G. A. (coords.). *Pobreza y Vulnerabilidad: debates y estudios contemporáneos en México*. Mexico, CIESAS, pp. 212–239.

García-Acosta, V. (2018b) Anthropology of Disasters. In: Callan, H. (ed.). *The International Encyclopedia of Anthropology*. New York, John Wiley & Sons, Ltd., pp. 1622–1629.

García-Acosta, V. (2014) Desastres históricos y secuelas fecundas. Discurso de ingreso a la Academia Mexicana de la Historia. In: *Memorias de la Academia Mexicana de la Historia. Correspondiente de la Real de Madrid*. Mexico, Academia Mexicana de la Historia, LV, pp. 65–91.

García-Acosta, V. (coord.). (2008) *Historia y desastres en América Latina*, vol. III. Mexico, LA RED-CIESAS.

García-Acosta, V. (2005) El riesgo como construcción social y la construcción social de riesgos. *Desacatos. Revista de Antropología Social*. 19, pp. 11–24.

García-Acosta, V. (2004) La perspectiva histórica en la Antropología del riesgo y del desastre. Acercamientos metodológicos. *Relaciones. Estudios de historia y sociedad.* XXV (97), pp. 123–142.

García Acosta, V. (coord.). (1997) *Historia y desastres en América Latina,* vol. II. Lima, LA RED-CIESAS-Tercer Mundo Editores.

García Acosta, V. (coord.). (1996) *Historia y desastres en América Latina,* vol. I. Bogota, LA RED-CIESAS-Tercer Mundo Editores.

Gellert, G. (2012) El cambio de paradigma: de la atención de desastres a la gestión del riesgo. *Boletín Científico Sapiens Research.* 2 (1), pp. 13–17.

Giesecke, A. & Silgado, E. (1981) *Terremotos en el Perú.* Lima, Ediciones Rickchay.

Haller, A. (2010) Yungay: Recent Tendencies and Spatial Perceptions in an Andean Risk Zone. *Espacio y desarrollo.* 22, pp. 65–75.

Hocquenghem, A. M. & Ortlieb, L. (1992) Eventos El Niño y lluvias anormales en la costa del Perú: siglos XVI-XIX. *Bulletin de l'Institut Français d'Études Andines.* 21 (1), pp. 197–278.

Hoffman, S. (1999a) The Worst of Times, the Best of Times: Toward a Model of Cultural Response to Disaster. In: Oliver-Smith, A. & Hoffman, S. (eds.). *The Angry Earth: Disaster in Anthropological Perspective.* New York and London, Routledge, pp. 134–155.

Hoffman, S. (1999b) After Atlas Shrugs: Cultural Change or Persistence After a Disaster. In: Oliver-Smith, A. & Hoffman, S. (eds.). *The Angry Earth: Disaster in Anthropological Perspective.* New York and London, Routledge, pp. 302–325.

Hoffman, S. (1998) Eve and Adam Among the Embers: Gender Patterns After the Oakland Berkeley Firestorm. In: Enarson, E. & Hern Morrow, B. (eds.). *The Gendered Terrain of Disasters: Through Women's Eyes.* Westport, Greenwood Publishing Group.

Hoffman, S. & Oliver-Smith, A. (eds.). (2002) *Catastrophe & Culture: The Anthropology of Disaster.* Santa Fe, Nuevo México, School of American Research.

Huertas, L. (2009) *Injurias del tiempo: desastres naturales en la historia del Perú.* Lima, Universidad Ricardo Palma, Editorial Universitaria.

Huertas, L. (1987) *Ecología e historia. Probanzas de indios y españoles referentes a las catastróficas lluvias de 1578, en los corregimientos de Trujillo y Saña. Francisco Alcócer, escribano receptor.* Chiclayo, CES Solidaridad.

Junquera, C. (2002) Antropología y desastres naturales: aportes y sugerencias factibles desde la investigación antropológica. *Espacio y Desarrollo.* 14, pp. 85–110.

Kámiche, J. & Pacheco, A. (2010) ¿Cuánto es afectado el consumo de los hogares cuando ocurre un desastre de origen natural? un análisis empírico para el Perú, 2004–2006. *Apuntes.* 67, pp. 67–107.

Lavell, A. (1993) Ciencias sociales y desastres naturales en América Latina: un encuentro inconcluso. *Revista EURE.* XXI (58), pp. 73–84.

López, M. (1999) La contribución de la Antropología al estudio de los desastres: el caso del Huracán Mitch en Honduras y Nicaragua. *Yaxkin.* XVIII, pp. 5–18.

Marino, A. (2015) *Estudio de los desastres socionaturales desde la psicología social: el caso Karl.* PhD in Psychology. Mexico, Universidad Veracruzana.

Mayer, E. (2009) *Cuentos feos de la reforma agraria peruana.* Lima, Instituto de Estudios Peruanos.

MINAM (Ministerio del Ambiente). (2016) *Historia ambiental del Perú. Siglos XVIII y XIX.* Lima, Ministerio del Ambiente.

MINAM (Ministerio del Ambiente). (2015) *Estrategia nacional ante el cambio climático.* Lima, Ministerio del Ambiente.

Montoya, R. (2016) Visiones del Perú en la Antropología peruana (1941–2015). *Investigaciones sociales*. 20 (37), pp. 15–30.

Montoya, R. (1970) *Algunas notas sobre el Callejón de Huaylas después de la tragedia.* Lima, DESCO.

Oliver-Smith, A. (2017) Adaptation, Vulnerability and Resilience: Contested Concepts in the Anthropology of Climate Change. In: Kopnina, H. & Shoreman-Ouimet, E. (eds.). *Routledge Handbook of Environmental Anthropology.* London, Routledge, pp. 206–218.

Oliver-Smith, A. (2016) La construcción social de los desastres: un reto para la antropología aplicada en el Perú andino. In: Ávila, J. & Bolton, R. (eds.). *Antropología aplicada en el Perú de hoy. Estudios de casos.* Lima, Colegio Nacional de Antropólogos del Perú.

Oliver-Smith, A. (2015a) Hazards and Disaster Research in Contemporary Anthropology. In: Wright, J. (ed.). *International Encyclopedia of the Social and Behavioral Sciences.* Amsterdam, Elsevier, 2nd edition, pp. 546–553.

Oliver-Smith, A. (2015b) *Me tocó documentar la reconstrucción del Yungay destruido* (Interview). Available from http://elcomercio.pe/peru/ancash/me-toco-documentar-reconstruccion-yungay-destruido-368275. Accessed November 2018.

Oliver-Smith, A. (2013) Disaster Risk Reduction and Climate Change Adaptation: The View from Applied Anthropology. 2013 Malinowski Award Lecture. *Human Organization.* 72 (4), pp. 275–282.

Oliver-Smith, A. (2002) El gran terremoto del Perú, 1970: el concepto de la vulnerabilidad y el estudio y la gestión de los desastres en América Latina. In: Lugo, J. & Inbar, M. (comp.). *Desastres naturales en América Latina.* Mexico, Fondo de Cultura Económica, pp. 147–160.

Oliver-Smith, A. (1997) El terremoto de 1746 en Lima: el modelo colonial, el desarrollo urbano y los peligros naturales. In: García-Acosta, V. (coord.). *Historia y desastres en América Latina.* Lima, LA RED-CIESAS-Tercer Mundo Editores, vol. II, pp. 133–161.

Oliver-Smith, A. (1986) *The Martyred City: Death and Rebirth in the Andes.* Alburquerque, University of New Mexico Press.

Oliver-Smith, A. (1979) The Yungay Avalanche of 1970: Anthropological Perspectives on Disaster and Social Change. *Disasters.* 3 (1), pp. 95–101.

Oliver-Smith, A. & Hoffman, S. (eds.). (1999) *The Angry Earth: Disaster in Anthropological Perspective.* New York and London, Routledge.

Ortiz, F. (1947) *El huracán: su mitología y sus símbolos.* Mexico, Fondo de Cultura Económica.

PCM. (2014) *Plan Nacional de Gestión del Riesgo de Desastres 2014–2021.* Lima, Presidencia del Consejo de Ministros (PCM), Secretaría de Gestión del Riesgo de Desastres.

Pérez-Mallaína, P. E. (2008) Las otras secuelas de una catástrofe natural. Tensiones sociales e ideológicas en Lima tras el terremoto de 1746. In: García-Acosta, V. (coord.). *Historia y desastres en América Latina.* México, CIESAS-La Red, vol. III, pp. 187–228.

Pérez-Mallaína, P. E. (2005) Las catástrofes naturales como instrumentos de observación social: el caso del terremoto de Lima en 1746. *Anuario de Estudios Americanos.* 62 (2), pp. 47–76.

Pérez-Mallaína, P. E. (2001) *Retrato de una ciudad en crisis. La sociedad limeña ante el movimiento sísmico de 1746.* Sevilla, CSIC, Pontificia Universidad Católica del Perú.

PNUD/PMA. (2010) *Mapa de Vulnerabilidad a la Desnutrición Crónica Infantil desde la Perspectiva de la Pobreza.* Lima, Programa de las Naciones Unidas para el Desarrollo (PNUD)/Programa Mundial de Alimentos (PMA), UN.

Quinn, W., Neal, V. & Antúnez de Mayolo, S. (1987) El Niño Occurrences Over the Past Four and a Half Centuries. *Journal of Geophysical Research.* 93 (Cl3), pp. 14449–14461.

Renique, J. (1992) Antropología e ideología. Notas sobre un artículo controvertido. *Debate Agrario*. 15, pp. 145–159.

Rivasplata, P. (2015) Protegiéndose del río Rímac: Los tajamares o muros de contención de Lima durante la colonia. *Investigaciones Sociales*. 19 (34), pp. 111–130.

Rosas, R. (ed.). (2001) El fenómeno "El Niño" en la Costa Norte del Perú a través de la historia; Perú-Ecuador, un espacio compartido. *I y II Jornadas de historia*. Piura, Facultad de Ciencias y Humanidades, Departamento de Humanidades, Universidad de Piura.

Seiner, L. (2011) *Historia de los sismos en el Perú: catálogo siglos XVIII-XIX*. Lima, Universidad de Lima.

Seiner, L. (2002) *Estudios de historia medioambiental: Perú, siglos XVI-XX*. Lima, Universidad de Lima.

Seiner, L. (2001) El fenómeno El Niño en el Perú: reflexiones desde la historia. *Debate agrario*. 33, pp. 1–18.

Silgado, E. (1978) *Historia de los sismos más notables ocurridos en el Perú (1513–1974)*. Lima, INGEMMET.

Sökefeld, M. (2013) Exploring the Link Between Natural Disasters and Politics: Case Studies from Peru and Pakistan. *Scrutiny. A Journal of International and Pakistan Studies*. 5, pp. 71–101.

Starn, O. (1992) Antropología andina, "andinismo" y Sendero Luminoso. *Allpanchis*. 39, pp. 15–71.

Taddei, R. (2014) Sobre a invisibilidade dos desastres na antropologia brasileira. In: Valencio, N. F. L. S. (ed.). *Waterlat-Gobacit Network Working Papers, Thematic Area Series Satad, TA8 – Water-Related Disasters*. 1 (1), pp. 30–42.

Torrico, G., Ortiz, S., Salamanca, L. A. & Quiroga, R. (2008) *Los enfoques teóricos del desastre y la gestión local del riesgo*. La Paz, NCCR, OXFAM, FUNDEPCO.

Urrutia, J. (1992) Comunidades campesinas y antropología: historia de un amor (casi) eterno. *Debate Agrario*. 14, pp. 1–16.

Vega-Centeno, M. (2011) Los terremotos, el crecimiento económico y el desarrollo. *Economía*. XXXIV (67), pp. 57–80.

Vega-Centeno, M. & Remenyi, M. A. (1984) Análisis económico de los terremotos: enfoque metodológico e implicaciones de política. *Economía*. VII (14), pp. 117–171.

Walker, C. H. (2012) *Colonialismo en ruinas. Lima frente al terremoto y tsunami de 1746*. Lima, Instituto de Estudios Peruanos.

Walker, C. H. (2004) Desde el terremoto a las bolas de fuego: Premoniciones conventuales sobre la destrucción de Lima en el siglo XVIII. *Relaciones. Estudios de historia y sociedad*. XXV (97), pp. 30–55.

Zilbert, L. (2008) Los desastres: ¿problemas no resueltos del desarrollo? In: Pradel, M. (coord.). *Territorio y naturaleza. Desarrollo en armonía. Serie: Perú hoy*. Lima, DESCO.

8 Anthropology of Socionatural Disasters in Uruguay

Javier Taks

Introduction

"En Uruguay no pasa nada". This saying, which can be translated as "nothing ever happens in Uruguay", is still used by youngsters and adults alike to explain the slow pace of social or technological change and the resulting sense of boredom. However, in the *"paisito"* – as Uruguayans like to call this 180,000 km² piece of land located between the two giants of South America, Brazil and Argentina – this idea that nothing really extraordinary happens there is linked to the idea that calamities of natural origin are rare. Until recently, phenomena such as floods, droughts, or windstorms, which do occur regularly, were not conceived as "disasters" but as typical features of the climate. In fact, there are instrumental and non-instrumental records of severe droughts and floods, and even earthquakes, and tsunamis in Uruguay from at least the 19th century that had a significant impact on everyday life.[1] What seems to happen in relation to disasters based on natural hazards is that they are embedded in the collective history, as fleeting exceptions or extreme manifestations of the normal, so as to cushion conflicts and diminish differences (Real de Azúa, 1984). As explained nearly a hundred years ago to potential settlers by a well-known promoter of agriculture referring to the regional climate, "[In Uruguay] what is abnormal is the norm!" (Boerger, 1928: p. 21).

Considering that local anthropologies are not alien to culture hegemony, it is not surprising to note that the anthropology of risk and disaster is just incipient. This chapter attempts to explore why social anthropology in Uruguay has given limited attention to disasters based on natural hazards, focusing on internal factors in the development of the discipline and its community of practice, since it was first institutionalized at the Universidad de la República (Republic University, Udelar) in 1976. The chapter starts with a brief history of academic anthropology in Uruguay, emphasizing how the nature-society relation was conceived both in social anthropology and archaeology. The thematic genealogies that influence the choice of objects of research and fields of intervention will be analyzed, considering the context of dematerialization of social anthropology during the height of cultural constructivism, which has gradually been qualified in the current second decade of the 21st century with the return of matter to the anthropological literature through the analysis, among others, of the environmental crisis.[2] As will become clear, this chapter embraces insights from my personal perspective as a main promoter in Uruguay of a line of research and training in the field of Ecological Anthropology and Environmental Anthropology.

In the second section I present a first overview of anthropological studies in the 21st century dealing with disasters based on natural hazards: droughts, floods, and severe storms, as well as tornadoes, examining how the phenomena was discussed in the context of the state's climate policy.[3] As discussed in Taks (2019), this governmental climate policy has gradually adopted global climate change as an umbrella narrative for human-nature relations, thus leading to a novel friction between imagined futures of normality and exceptionality. This friction, in turn, has consequences for how disasters based on natural hazards are conceived and faced.

The selection of texts resulted, primarily, from a search in two academic journals published in Uruguay, namely, the Udelar's *Anuario de Antropología Social y Cultural* (2000–2015), renamed *Revista Uruguaya de Antropología y Etnografía* in 2015, and the more recent *Revista Trama* (2009–) edited by a professional body called *Asociación de Antropología Social del Uruguay (AUAS)*. In addition, colleagues in the Department of Social and Cultural Anthropology,[4] and in the Department of Archeology of Udelar, provided names of researchers and suggested papers published in international journals and presented at academic events, which were also included in the overview where relevant. The criterion for

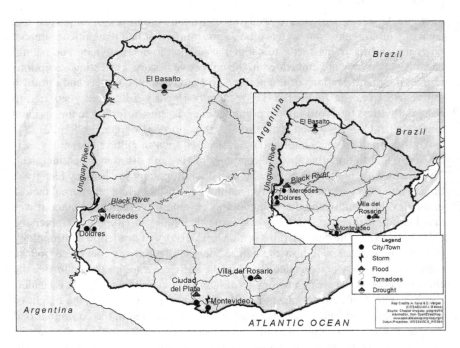

Map 8.1 Map Uruguay. Case studies and main areas mentioned

determining whether the work was anthropological was whether its author had an undergraduate or postgraduate degree in social anthropology either.

Anthropology in Uruguay through nature-society relationship

A comprehensive history of Uruguayan Anthropology is still to be written,[5] as well as an analysis of its changing conceptions on the relations between nature, society and culture, a topology where, in principle, we can locate the study of socionatural disasters. In this section, I propose a first outline for such a task centered on the development of social anthropology, though including information from local archeology where relevant, and claim that greater dialog between these branches of anthropology is needed, for understanding the formation of past and present socionatures.

The recent history of social anthropology in Uruguay can be divided into four periods, which I preliminary call: pioneering stage, stage of dictatorial exotizising, stage of democratic transition and stage of consolidation and affirmation of the discipline. In what follows each stage will be described, focusing on how the relations between nature and society were conceived and, particularly, the place reserved, if any, to the treatment of disasters based on natural risks.[6]

Pioneering stage

Guigou (2016: pp. 62–63) traces back the academic foundation of a "native" social anthropology between the years 1964 and 1968, when the Brazilian anthropologist Darcy Ribeiro, who was exiled in Uruguay, read courses in Cultural Anthropology at the Faculty of Humanities and Sciences (FHC) of Udelar. Renzo Pi Hugarte, anthropologist, friend, collaborator, and translator of Ribeiro is considered his follower (Ribeiro Coelho, 2003: p. 45), lecturing on the same subject in the Faculty of Law and Social Sciences in the same University. Meanwhile, between 1967 and 1972, the anthropologist Daniel Vidart lectured on Cultural Anthropology at the then School of Social Welfare, also at Udelar. These first attempts to develop social anthropology as an academic discipline ended with the coup d'état and the civil-military dictatorship (1973–1985) which intervened Udelar and facilitated the exile of the pioneers, Vidart and Pi Hugarte. Before exile, they were concerned with the contribution of indigenous peoples and the legacy of immigrants to the construction of the national identity (Vidart & Pi Hugarte, 1969; Pi Hugarte, 1969). Of the scholars working in this early period, it was Vidart who engaged directly with the subject that concerns us here. He presented an early relational model between nature and culture through his description of the rural and urban landscapes and the emergent subjectivities that he called "human types" of the countryside and the city. His writings, between descriptive and poetic, constitute a good precedent to the more recent phenomenological approaches to the perception of the environment among livestock herders (De Torres & Piñeiro, 2013).

Cabrera (2011) places also in the 1960s the beginnings of archeology as a scholarly endeavor in the country, pointing to the work of Antonio Taddei, considered by many to be the father of Uruguayan Archeology. Taddei founded the Center for Archaeological Studies in 1969 and played a leading role in the organization of various meetings of archeologists at national level. Within the discipline of archeology, the relationship between society and nature was based on a hypothesis of conditionality of the natural forces towards human social development. For example, in the description of Taddei's classic study of the archaeological site of arroyo Catalan, Cabrera (2011: pp. 85–86) refers to the collaborative work of geologists and archaeologists who concluded that cultural development and peopling by *catalaense* prehistoric hunters can be strongly associated with a dry period prior to the optimum climaticum, i.e., between 8,000 and 9,000 BP, which is manifested in the geological and hydrogeological regional record.

Stage of dictatorial exotizising

As mentioned previously, the pioneering stage effectively ended in 1973 with the military intervention of the University. Yet the year 1976 is an important landmark in this brief recount because, three years after the coup d'état, the first Bachelor's Degree in Anthropological Sciences (Licenciatura en Ciencias Antropológicas, LCA), with its two branches – Archeology and Prehistory and Cultural Anthropology – was launched in the FHC of Udelar.[7] However, Guigou (2016: p. 62) argues that there were strong mechanisms of censorship restricting the theoretical discussions and studies that could offer a critical understanding of the harsh reality of the dictatorship. He adds that folklorism, a focus on global rarity, and the teaching of neutral theories of culture and society were dominant, concluding boldly that the period between 1976 and 1985 left no significant legacy for national social anthropology.

The development of archeology was significantly affected by the international archaeological rescue mission carried out between 1976 and 1983, prior to the construction of the binational hydroelectric dam of Salto Grande, on the Uruguay River. Under the auspices of Unesco and the Ministry of Education, under the leadership of the French archaeologist A. Laming-Emperaire (followed by archaeologist Niède Guidon), many of the students of the new Department of Anthropological Sciences made their first excavations and laboratory work. For our purposes, it is important to note that the construction of the Salto Grande dam was the first large development project where an environmental impact study was applied, influenced by the sustainable development paradigm that began to be hegemonic in the Euro-Atlantic region (CTMSG, 1977).

Stage of democratic transition

On 1 March 1985, a new civil president took office after quasi-free general elections, marking the beginning of the official democratic transition for many state institutions, including Udelar. Teachers coming from abroad (politically exiled or not), as well as young and mature students formed the critical mass for the renewal of the Bachelor's Degree. Vidart and Pi Hugarte were part of that relaunching, with Vidart becoming the first director of the LCA in 1985 until his resignation in 1988.[8] Pi Hugarte followed him as Director during several years. Overall, the last half of the decade of the 1980s was devoted to theoretical discussions about the development of curricula that resulted in the 1991 Program, which was to become the guiding chart for academic training in Anthropology for the next 20 years in their two classic orientations.[9]

The influence of symbolic anthropology and the postmodern turn is important at this stage. The theme of national cultural identity was much debated in the context of the ideological reconstruction of the Uruguayan exceptionality as a unitary and homogeneous nation. Anthropology in all its branches was to make a critique of the monocultural nationalism promoted by the military regime, calling for the recognition of cultural diversity from prehistoric times to contemporaneous subnational regions and urban settings, not to mention seminal research on genetic markers among living populations. In social anthropology, discourse and representation became the main focus of attention, replacing technology, history, and social structure.

Interestingly, there was also at this stage interest on material processes that addressed the dialectical relationship between nature and culture, although this was more in discussion in the classroom rather than in empirical research. For instance, environmental issues were first addressed in Vidart's lectures on Human Ecology, based on his book *Environmental Philosophy* (Vidart, 1997). The study of natural phenomena and their socioeconomic effects was also central in the work and lectures given by Guillermo Foladori. Foladori was trained in anthropology and economics in Mexico in the late 1970s; he joined the Department of Anthropological Sciences of the FHC in the year 1986 and worked there up to 1991. The relationship between nature and society was discussed in his course on Economic Anthropology, particularly in his presentation of the Marxist labor theory of value and the associated concept of subsumption of labor to capital. This concept had been developed in his dissertation on the partial proletarization of peasants in Mexico (Foladori, 1985) determined by a greater dependance of capitalist agricultural production on natural rhythms of plant and animals' growth. Moreover, his course's reading list included a classic of Marxist economic history, namely Witold Kula, employed to discuss methodological matters after the mandatory reading of a chapter entitled "Man's dependence upon nature" and particularly its sections devoted to "Natural disasters in history", "Historical studies on climate", and "Natural disasters and class struggle" (Kula, 1977). After he left the Department of Anthropological Sciences, Foladori continued his work on methodological aspects for the study of environmental perception in the Department of Sociology at the recently created Faculty of Social Sciences (Foladori et al., 1996). That research project resulted in several publications in the field of environmental studies and, more importantly, helped to establish a multidisciplinary team of researchers that expanded his work on sustainability from the standpoint of anthropology and other social sciences (Foladori & Taks, 2004; Foladori et al., 2005a, 2005b). These were the first steps towards what would become a line of research in ecological and environmental anthropology.

Stage of consolidation and affirmation

Since the mid-1990s and with greater clarity in the 21st century, social anthropology consolidated its place in Uruguayan academic institutions. According to Guigou (2016), indicators of such consolidation are these: the launching of the first Master's degree in Anthropology in the Facultad de Humanidades y Educación (Faculty of Humanities and Education, FHCE); the ratification of the Department of Social Anthropology within the Instituto de Ciencias Antropológicas (Institute of Anthropological Sciences), along with the Department of Archaeology and the Department of Biological Anthropology; the increasing employability of social anthropologists in the university, state agencies and in the private sector; the existence of two specialized journals; and the foundation in 2005 of the already mentioned professional association AUAS. From a theoretical point of view, Guigou identifies an ethnographic and post-Eurocentric turn, which translates itself in the question "What does it mean to think from the South?", which was the topic of the XI Meeting of the Anthropology of Mercosur (RAM) organized in Uruguay in 2015.[10] Rostagnol (2016) agrees with the general image of this last stage as a moment of deconstructing both the culturalist and structuralist thought of Euro-American roots. She inscribes 21st century social anthropology in Uruguay within the so-called anthropologies of the world, a movement and plea for de-Westernizing and decolonizing the discipline. She emphasizes the following main features of local anthropology, shared with other Latin American schools: the de-familiarization of day-to-day practices through anthropologies made at home; the articulation of micro and macro perspectives through the depiction of contexts for particular meanings; the search for the Other and social heterogeneity in the current "multiverse"; a commitment to social reform and change towards "increasing equity and the realization of rights"; and the practice of "Latin American anthropological ethnography", which seems to put together the study of otherness with political and cultural activism, favoring an agenda of collaborative research (Rostagnol, 2016: pp. 42–43).

During the first decade of the new century, and clearly apart from the aforementioned post-Eurocentric turn, Foladori (2005a, 2005b) addressed the theme of natural disasters to illustrate his critique of the role of science in the separation between climate change and nature, on the one hand, and social relations on the other. He pointed out that, in order to understand the differences between countries and between social classes in facing the impacts of natural forces, we need to study social vulnerability (produced primarily by the dynamics of capital accumulation and its inescapable social differentiation) and not so much the extreme climatic phenomena in their physical characteristics. In that way, he wanted to mobilize scientific advancement for the improvement of common peoples' life. In his words: "For the vast majority of the population it is much more important to reduce vulnerability against extreme events rather than to reach absolute accuracy on the prognosis and measurement of the intensity of the phenomena" (Foladori, 2005a: p. 56). The intellectual heritage of Foladori for local environmental anthropologies lies mainly in his choice of a "critical humanist anthropocentrism"

(Foladori, 2005c) standpoint wherefrom to understand environmental history as a coevolution of human society and nature and, simultaneously, the recognition of human society as internally divided into large antagonistic groups, each of them with a differential responsibility in how the biophysical environment is transformed according to an unequal access and control of the production processes. Furthermore, the increasing gap between social groups in terms of decision-making capacity and the practical transformation of the environment goes hand in hand with an increasing alienation of human society and its processes of becoming with respect to nature.[11]

Following Foladori's insights, Taks (2001) conducted his doctoral thesis on the transformation of environmental perceptions among producers of milk, called *tamberos*, during a techno-productive conversion in the area of Villa del Rosario, Province of Lavalleja, in Eastern Uruguay (see Map 8.1). This work was also strongly influenced by Tim Ingold's critique of cognitivism and culturalism in addressing traditional environmental knowledge (Ingold, 2000). In his analysis of people's ways of resolution of the tension between task-oriented time and commodity time, Taks demonstrates how an alienated perception of the weather, expressed in a less trust in their own readings of the landscape, and a more hostile representation of weather and climate *vis-à-vis* their households' needs, were linked primarily to farmers' financial indebtedness to loan lenders (banks, dairy-cooperatives), but also to a trend towards productive specialization and agricultural intensification, along with the threat from climate change communicated through the media (Taks, 2001: pp. 259–269; see also Taks, 2000).

Taks and Foladori's ethnographic observations and anthropological reflections on the perception of weather and climate led to a series of interdisciplinary meetings within the framework of Udelar's Enviromental Network (RETEMA in its Spanish acronym),[12] particularly with climate scientists and agronomists, where anthropological knowledge was called upon for contributing to resolve a paradox, already outlined in Taks (2001), between the results of scientific analysis of historical climate series and practical knowledge of agricultural producers and extensionists. In the following section, I offer a preliminary review of recent anthropological papers and theses that directly or indirectly follow on from the debate on climate change that emerged from those meetings, and continue the path of the earlier depicted environmental anthropology.

Anthropology of extreme weather events

For the last decade and a half, extreme weather events, mainly drought,[13] provoked dialog and interdisciplinary discussions, including anthropologists, under the umbrella of the debate on climate change (Bidegain et al., 2012; Astigarraga et al., 2013; Bartaburu et al., 2013). As argued in Taks (2001), whilst climate scientists conclude that there is no statistical novelty in climate variability according to the last 100 years' records, inhabitants of rural areas speak of recent significant changes in the cycles of plants, animals, and in their own bodies associated with different weather conditions. The aim of most anthropological research, therefore, has been to reembed climate and extreme weather events into the local and national social matrix, understanding the relations between hegemonic and subaltern voices, as well as the analysis of the material and political effects of a particular form of conceiving nature.

Anthropologists De Torres and Taks (De Torres et al., 2007) approach the "social use of climate" among livestock herders during agronomic draughts. As part of a research team on agrometeorology for a regional comparative study funded by the Inter-American Institute for Global Change Research (IAI), these anthropologists argue that the externalization of nature, in this case the understanding of drought as a natural phenomenon without possibility of human intervention in its course, was closely linked with the interests of certain social groups in search of economic benefits and co-optation of state action. Such an alienated view of natural phenomena neglects to account for the structural differences between livestock producers to prevent, mitigate, or suffer water scarcity. Following Taddei's methodology (2005), this study was the first qualitative exercise on collecting memories of droughts for an analysis of the material and discursive resources mobilized strategically, individually, and collectively to face them. Moreover, it revealed the differential handling of climate information according to productive economic types of cattle ranchers. After this first study, other research teams worked with testimonies and press archives to give an account of continuities and change on how droughts were objects of social struggle and claims at national level, comparing the draughts in the years 1943, 1988/89, and 2009 (Bidegain et al., 2012).

In addition, De Torres and Piñeiro (2013) carried out an in-depth ethnographic study of the relationship between livestock herders and droughts. They called farmers in the north of the country in a semiarid area known as *El Basalto* the "guardians of the meadow". Inspired by Ingold's (2012) concept of wayfaring as the primary form of environmental perception, they explore cattle herders' perception of the atmosphere and the environment during their journeys on horseback or by foot through the fields and meadows. According to their findings, herders engage with their surroundings basically as a bodily experience and their sense of smell, strength, and directions of winds, among other signs, warn them about extreme weather and climate events.

Beyond everyday perception, drought as disruption cannot be understood without acknowledging historical changes in the form of production, for instance the need to increase the number of heads of cattle per unit of land to satisfy market demands. Vulnerability in relation to draught is multidimensional, involving technical, institutional, and knowledge factors (Bidegain et al., 2012: p. 101). Draught is a type of disaster that combines a gradual transformation with moments of strong impact when certain thresholds, both physical and social, have been exceeded. In those disruption moments, producers' demands to the state play a fundamental part. Diego Thompson (2016) discusses the problem of governance, territorial decentralization, and adaptation to climatic and environmental challenges in the southwestern part of the country. Here, many conflicts arise around drought and the definition of climate emergency as a result of the dependance of local communities on centralized institutions for access to economic, financial, physical, and informational resources, which are key for adaptation to environmental stress. On the other hand, Thompson also notes that the actors involved rarely objectify the prevention of disasters and only act when the disaster is already unpacked (Thompson, 2016).

The opposite of agronomic droughts in the Uruguayan collective imagination of disasters are floods. Although floods have an impact on rural activities, it is in urban environments where flooding have been conceptually turned into disasters and sources of risks, particularly since the mid-20th century. The flood of the year 1959 is an unavoidable milestone for understanding the relationship between social organization and meteorological events, attracting much greater visibility for urban problems and the exposure of its inhabitants to the risk of sudden flooding. Flooding has recently attracted greater interest on the part of hydraulic engineers, urban planners, and technicians along with the creation in 2009 of the National Emergency System, a network of state institutions that focus on relief in the aftermath of calamities.

The occupation of floodplains for residence has become a widespread phenomenon from the 1980s, mainly on the west coast of the country. The majority of occupied land are state-owned terrains located on the banks of rivers and streams, where people displaced from the consolidated city have seen possibilities of occupation and housing, usually in very precarious conditions, and often encouraged by local authorities or entrepreneurs in the undercover economy. The result is the development of what some call "flood culture", a pattern of temporary displacement according to the rhythms of rainfall without abandoning flooding plots permanently. State authorities and technicians, as well as scholars involved in urban planning, adopting a hygienist and/or environmentalist approach, struggle to contain that flood culture which, according to their view, is responsible for both the displacement of some 70 thousand people since the beginning of this century and for other intangible problems (Piperno et al., 2015: p. 542).

The social vulnerability of flood-prone zones converts the phenomenon of urban flooding in disaster. However, urban flooding was treated until recently just as a contingent and extraordinary phenomenon, according to the study of Contreras about flooding in the city of Mercedes, the capital town of the Province of Soriano on the southern margins of the Black River. She points out:

> Once the flooding ends as a chaotic moment, everything returns to its previous order without contestation. On the contrary, the flood seems to reaffirm the two great pillars on which is based our society: its internal differences following a class structure and the division between nature and culture.
>
> (Contreras, 2012: p. 53)

According to Contreras' ethnography, class structure is revealed in the opinion of members of the upper and middle classes, as well as in the attitude of state officials, that floods affect poor people because the latter have developed an "evacuation industry", taking advantage of state welfare policies. On the other hand, the dichotomy between nature and culture manifests itself in the dominant language that floods are a result of natural phenomena related, primarily, to extraordinary rainfall in the watershed, outside of any peoples' control. Moreover, if any anthropic actions are seen as influencing such events as floods, they are associated to global processes of great abstraction, as for instance the increase of the greenhouse effect that leads to the acceleration of the hydrologic cycle, which in turn results in more intense rainfall. In other words, there is a naturalization of the anthropic action when passing from the local to the global, from where it comes back as an external force in the local explanation of the sources of rain and flooding.

The limited cases of anthropological research on flooding (Contreras, 2012; see also Corena, 2018) seem to contest the unilateral gaze of urban planning, proposing that flood culture has its positive aspects, such as tuning social and natural rhythms, which might give rise to collective measures to improve local resilience, countering the alienating effects of relocating inhabitants outside their territories. The discipline also seems to assume that people are living and facing in the best way possible the gradual transformation of weather and climate, embedding them in social time and into their taskscape.

Obviously, each extreme event acquires specific meanings according to the time and place of occurrence. In relation to severe storms, Taks (2019), as part of a research team, studied one of the most salient political-climatic events of this century in the country. They refer to the storm on 23 August 2005, technically defined as an "extra-tropical cyclone", that the public weather services did not foresee and caused human and animal deaths, as well as great material damage, especially in the capital, Montevideo. The researchers analyzed media reporting and personal accounts around this event, and came up with interesting insights on the situated science of climate scientists. One of the main findings was that climate scientists (affiliated mainly to Udelar) propose an understanding of climate as a neutral object, while end-users of their information and communication live climatic phenomena involved in subjective landscapes of stories, traversed by feelings that recreate that particular time, bringing to the fore their embodied memories. Yet, another group of actors, nonexperts like decision-makers, producers, and agricultural technicians, journalists, among others, cannot recognize uncertainty as part of reality but just as a byproduct of poor science and limited knowledge. Their demand for certainties (Taddei, 2012) is something that challenges meteorologists and climate scientists in their expertise for the development of forecasts. Yet, more importantly, this unilateral claim for better forecasts hides the historical construction of vulnerability and risk in certain territories and between social groups with limited access to material and symbolic resources (Murgida & Radovich, in this volume). Another important conclusion of the study is that that storm of 2005 acted as a catalyst for public policy in relation to climate change: it contributed to the creation of the National System of Response to Climate Change in 2009, it accelerated a launching of a Bachelor's Degree on Atmosphere Sciences at Udelar, and it made visible the material poverty and rather limited legitimacy of the former meteorology institute attached to the Ministry of Defense. Not surprisingly, it was subsequently converted into the Uruguayan Institute of Meteorology

(INUMET), under the responsibility of Ministry of Housing, Land Management and Environment.

More recently, in April 2016, another event, this time a windstorm, made visible urban vulnerability. A tornado of magnitude 3 on the Fujita-Pearson Scale destroyed the center of the city of Dolores (20,000 inhabitants) in the Province of Soriano, causing one human death and the loss of hundreds of homes as well as provoking severe damage to public and private buildings. The shock has left psychological scars, while at the same time reviving the concern about the increase in amount and intensity of this type of event in the future. Climate experts pointed out that that event was not unexpected from a scientific point of view because it is well known that the west of the country is a corridor of tornadoes of various magnitudes. Yet, beyond their opinion, it had never happened that a tornado had passed through the heart of a medium-scale city, which accelerated – along with the aforementioned coastal flooding – the start of projects of urban adaptation to climate change supported by international and global funds.[14] Thompson and López Barrera (2018) have recently explored the way in which the community of Dolores faced the challenges post-tornado. They highlight that a community that had not mobilized its social capital to face longer-term environmental problems, like atmospheric and water pollution as a result of the intensification of agriculture and agro-industry in its hinterland, nevertheless activated multilevel governance and collective action to deal with the unforeseen emergencies after the tornado and reconstruct the impacted neighborhoods. Unfortunately, Thompson and López Barrera argue that the way reconstruction took place reproduces multiple social and material vulnerabilities in the medium and long term in front of the same risks based on meteorological and climate hazards. The ignorance, negligence, or oblivion of risks in gradual processes of environmental degradation continues once the emergency has passed, in a similar vein to what has already been discussed previously in relation to flooding in nearby Mercedes (Contreras, 2012). According to Thompson and López Barrera (2018), perception of background risk in the case of Dolores is veiled by the image of collective economic wellbeing beyond social differences in the accumulation of wealth.

Generally speaking, the greater visibility of extreme events and its connection to the dominant narrative of global climate change has provided an opportunity for the state, and in a lesser extent for private actors, to view the country as part of a global humanity at risk against external nature. Nevertheless, at the same time, this view of a homogenized human society against nature needs to be officially qualified in the name of a self-identification of Uruguay as a member of a weak portion of that same humanity, victim of the global climate chaos set up by the more developed and powerful nations. Therefore, the adoption of the narrative of climate change helps to unite the nation against both nature and an asymmetric world, whilst domestically there appears to be a fragmentation of the political use of climate to push reforms in diverse sectors (energy, agriculture irrigation, transport), which depend on access to national and, mainly, international funding (Taks, 2019).

To sum up this section, this review of incipient anthropological studies of disasters based on natural hazards associated to meteorological phenomena, has shown first – following the classification of differing anthropologies of disaster and risk proposed by Virginia García-Acosta (2018)[15] – that, in Uruguay, the anthropological study of disasters and the perception of risk has been gradually made relevant from a political-economic and environmental point of view.

Second, there is an influence from historical materialism which proposes a coevolution of society and nature as a way of understanding the limits of the divide between these two domains of reality, without necessarily moving into the ontological turn and the proposition of symmetrical anthropology. Local studies show a primary concern with the analysis of relations between humans to understand the occurrence of certain socionatures, without neglecting the reality of extreme weather and climate events in human social relations. In this regard, the concept of disastrous landscape, as exposed by Camargo in this volume, might be useful in orienting anthropological research towards a more direct engagement with natural sciences knowledge, since the disaster causes a dramatic and measurable material transformation of the environment which in turn provokes ethnographic revelations for understanding the reconfiguration and restructuring of things and elements, both human and non-human, that mutually constitute these affected landscapes.

In terms of methodology, the works reviewed include the tracing of historic weather events, the collection and analysis of personal and collective memories, the study of written and audio-visual records, and the analysis of discursive and material resources mobilized by particular social groups to prevent or overcome disasters based on natural hazards. Implicitly acknowledged throughout this work is the idea emphasized by García-Acosta (2015) that what is important is to analyze the historical construction of vulnerability that enables a disaster to happen.

In one way or another, most of the studies reviewed here seek to politicize climate change, addressing the relationships between people and extreme events, valuing their knowledge in front of expert and technical knowledge, supporting and making public peoples' unique stories. This politicization should reach all discourses on climate and weather, destabilizing hegemonic discourses that, in the name of an idealized balance and harmony within society, normalize extreme events without firmly addressing the economic, political, and environmental vulnerabilities at the root of most socionatural disasters.[16]

Final remarks

Disasters based on natural hazards and disrupting everyday life have gained greater visibility in Uruguay, where they have become issues of political and academic agenda, as well as everyday topics of conversation.[17] The state, scientists, and the media have been educating our attention (Ingold, 2018: p. 20). In the 21st century, "natural disasters" are seen primarily as extreme weather events, and, more ambiguously, as the degradation of aquatic ecosystems. Before the Earth Summit of United Nations in 1992, meteorological phenomena were conceived as part of the climate features in the country (OPP/BID/OEA, 1992); however, since then, meteorological phenomena are increasingly contextualized as manifestations of global climate change, proposing a close relationship between climate change and extreme events (CLAES/PNUMA/DINAMA, 2008; Giménez et al., 2009), which can be clearly associated with the institutionalization within the state of global climate change as a narrative originated in the rhetoric and action of international agencies and epistemic communities (Conca, 2006; García Cartagena & Taks, 2015).

While in Uruguay it is still possible to find narratives that identify the causes of disasters exclusively in the domain of nature, that is, within the sphere of reality that escapes human control according to the modernist naturalist scheme (Descola, 1996), there is a growing collective consciousness that there are no "natural disasters" *per se* but rather climate and geological events with destroying capacity, which are simultaneously natural and social phenomena, where ideological aspects are increasingly recognized as inherent to the event itself.

I have contextualized the study of disasters based on natural risks within the social anthropology literature relating to the relationship between society and nature. The works of a small community of professional anthropologists, formed mainly after the democratic transition in the late 1980s, whose focus of attention was mainly the analysis of human relationships and intersubjective symbolic construction of reality, paved the way for the more recent discussions, which integrate the coevolution of these same social and cultural processes with the material realm of nature and its feedback effects on social relations and collective imageries.

Outside the academic literature, I noted the very lately acceptance by the Uruguayan government and other influential actors in public policy, that every disaster is better prevented if particular multilevel social structure and agencies are taken into consideration. Social sciences and humanities, until recently, seemed to have little to contribute to the diagnostics and techno-engineering solutions for drought, floods, storms, and the growing ecological degradation. This chapter has made a case for the development of this field of study and intervention in social anthropology in Uruguay. Before finishing, however, I would like to suggest a couple of questions that could be worth responding to before embarking on the task.

To begin with, can social anthropology respond to complex issues that emerge from the analysis and intervention on risk and disasters, independently of its sister disciplines of archeology, linguistics, and/or biological anthropology? Surely, it can give very good partial answers, but not holistic ones. Oliver-Smith and Hoffman have already suggested that, in disaster research, anthropology has a worthy opportunity "to amalgamate past and current cultural, ecological, and political-economic investigations, along with archaeological, historical, demographic, and certain biological and medical concerns" (Oliver-Smith & Hoffman, 2002: p. 6). For this reason, the fraternal dialog between anthropological subdisciplines would have to be strengthened if we want to establish a fertile critical materialism (Lettow, 2016) to propose and promote sustainable futures, which, among other things, might prevent natural disasters through the reduction of social vulnerability and exposure to risk.

Anthropology does not conceive just one but multiple and different ways of understanding and use risk. Since the mid-20th century, Uruguay was considered a homogeneous country in terms of cultural traditions, with a single climate and a geology of "gently rolling prairies", as children would learn at school. Globalization, understood as a process of the production of "uneven geographical developments" (Harvey, 2000: p. 94) has meant a change in the understanding of national development, where diversity, inequity and inequality in the territory become central, and closely associated with the various structures of risk. Therefore, shall social anthropology contribute to building a culture of risk prevention at the national level as suggested by some consulting work on public policies (ONU, 2015)?

From this, there follows another question. There is a contesting attitude and practice of many vulnerable groups against official policies and programs dealing with risk and disasters. How to explain this? There are at least two paths. One that states that there are diverse and frictional ontologies at play. And the other that it is the performance of adaptive preferences by disempowered people to confront inequities and inequalities. In this regard, two proposals are made. First, it seems worth studying the processes of production of hegemony in the definition of those risks that can be identified and those who cannot. Second, it seems necessary to develop criteria for classifying social vulnerabilities to allow different agents and organizations to understand the reproductivist or transformative nature of various solutions for preventing and dealing with disasters. For example, if overcrowded dwellings and lighter building materials among the poor produce vulnerability to extreme events, then the relocation to precarious housing designed incoherently with respect to family structures of those who are going to inhabit those dwelling maintains vulnerability and increases the certainty that there will be more disasters to come (Thompson & López Barrera, 2018).

Finally, how can anthropology relate and contribute to public policies regarding disaster prevention and relief? A comprehensive anthropology of vulnerability and disasters should avoid the fragmentation of socionatures resulting from the fragmented structure of the state, for example, droughts for agricultural production, floods for urban planning, and severe storms for national defense and insurance policies. On the contrary, anthropology in Uruguay and elsewhere should try to understand disasters, following Oliver-Smith (2017), as an arena of transversal disputes, embedded in a social, cultural, and multilevel territorial matrix, from households to the GEF-World Bank headquarters.

Notes

1 Seismographs have been installed in the country since 2013, helping to promote the idea that earthquakes were also becoming a larger territorial threat than previously believed.
2 Gender studies and analysis of women's bodies (Rostagnol, 2003), as well as the study of unequal urbanizations (Abin Gayoso, 2014), can be pointed out as contributing to the return of the matter to local social anthropology.
3 For reasons of space, the review does not include pioneering local studies in social anthropology centered on the social construction of risk in human health (Romero, 2010, 2014), addictions (Albano et al., 2014), urban mobility (Rossal & Fraiman, 2007), or industrial development (Renfrew, 2018).
4 I am particularly grateful to Victoria Evia from the Department of Social Anthropology for her dedicated reading and useful recommendations. Also, I acknowledge Gaby Saldanha for her English review and Anaclara Viera for a preliminary map design.

5 In 2018, for the first time in the Bachelor's Degree in Anthropological Sciences, an optional course entitled Social Anthropology in Uruguay was offered, taught by Dr. Susana Rostagnol.

6 An initial "naturalist" or "collecting" stage, extended from the late 19th century until the mid-20th century, could be added. It was the work of people (usually men) with an integral interest in archeology, linguistics, and ethnology (Cabrera, 2011). Their efforts were devoted to classifying ethnohistorical and prehistoric cultures in the territory of Uruguay, and to articulating local prehistory and history into a universal evolutionary model.

7 The Department of Anthropological Sciences was founded the previous year, in 1975. Its first director was the Argentinian archaeologist Antonio G. Austral.

8 Daniel Vidart's participation in Udelar was too short. After he left the LCA he lectured in a postgraduate degree in Education Sciences between 1990 and 1991. Later, in 2003, he was involved as a member of the UNESCO Chair in Human Rights of the Udelar for a couple of years.

9 A new curriculum was adopted in 2014. Available from www.fhuce.edu.uy/images/comunicacion/destacados/2014/marzo/planes/Nuevo%20Plan%20Antropologia.pdf Accessed 17 March 2019.

10 The two previous Meetings of Anthropology of Mercosur organized in Uruguay had as their central themes nothing comparable in terms of disciplinary self-reference: VI RAM, 2005, Montevideo, "Identities, fragmentation and diversity"; II RAM, 1997, Piriápolis, "Cultural borders and citizenship".

11 Following this line of thought, I have included some topics of Environmental Anthropology in the course of economic and political anthropology since 2001.

12 RETEMA was created in 2001, when the Udelar became aware that it was necessary to create spaces of multi- and interdisciplinary work to address objects of study, training, and outreach for which isolated disciplines show important limitations. The format of network (more than centralized interdisciplinary centers) was the challenge at that moment. Today the RETEMA is defined as a collaborative space of work on environmental issues, multidisciplinary and interdisciplinary, where researchers and lecturers seek to overcome the obstacles of the specialized training and institutional fragmentation for a better approach to the complexity of environmental education, the analysis of environmental conflicts and the making of relevant university extensionism. Available from http://udelar.edu.uy/retema Accessed 17 March 2019.

13 The extreme events recorded by the national historiography are from the 19th century and, together with collective memory, indicate that "drought" is the greatest protagonist. This seems logical in a country of extensive cattle herding, too dependent on pastures, which in turn are strongly dependent on soil moisture, precipitation, and temperatures, as well as sources of water for watering livestock (Jacob, 1969).

14 For example, the elaboration of the National Plan for Climate Change Adaptation in cities and infrastructures, known as NAP cities, with funds from a grant from the Preparatory Program of the Green Climate Fund agreed at the Paris Summit, managed by UNDP of the United Nations.

15 According to García-Acosta (2018) the other two topical areas and major approaches in disaster anthropology are a behavioral and organizational response approach and a social change approach.

16 Uruguay seems similar to Brazil in this strategy of normalizing the exceptional, in part for the same reason pointed out by Taddei (2017) that the hegemonic production of information about disasters happens in the main capital city away from the locations where the strongest effects of socionatural disasters are directly lived.

17 A recent United Nations mission to assess the state of reduction of disaster risk in Uruguay (ONU, 2015) listed the following existing disasters: regression of the coastline, floods, thunderstorms, degradation of environmental quality (mainly water), *dengue*, *chikunguya*, forest fires, winds, hail, and drought. Also mentioned were radiological

events and transport of industrial hazardous waste. From 2013, the degradation of water quality in the country is a subject of debate, where the natural and human-induced extreme events such as algal blooms are disputed among researchers, politicians, and communicators. (Thompson, 2018; Alonso et al., 2019).

References

Abin Gayoso, E. (2014) Por el derecho de los vecinos a vivir en su barrio: Cooperativa de vivienda en Ciudad Vieja de Montevideo. *Revista Trama.* 5, pp. 61–75.

Albano, G., Castelli, L., Martínez, E. & Rossal, M. (2014) Reducción de riesgos y daños del uso de pasta base de cocaína en Malvín Norte. *Integralidad sobre ruedas.* III (3), pp. 14–20.

Alonso, J., Quintans, F., Taks, J. & Conde, D. (2019) Water Quality in Uruguay: Current status and challenges. In: Vammen, K., Vaux, H. & de la Cruz Molina, A. (eds.). *Water Quality in the Americas. Risks and Opportunities.* Mexico, IANAS.

Astigarraga, L., Cruz, G., Caorsi, M. L., Taks, J., Cobas, P., Mondelli, M. & Picasso, V. (2013) *Sensibilidad y capacidad adaptativa de la lechería frente al cambio climático.* Vol. IV of *Clima de cambios: Nuevos desafíos de adaptación en Uruguay.* Montevideo, MGAP-FAO.

Bartaburu, D., Morales, H., Dieguez, F., Lizarralde, C., Quiñones, A., Pereira, M., Molina, C., Montes, E., Modernel, P., Taks, J., De Torres, M. F., Cobas, P., Mondelli, M., Terra, R., Cruz, G., Astigarraga, L. & Picasso, V. (2013) *Sensibilidad y capacidad adaptativa de la ganadería frente al cambio climático.* Vol. III of *Clima de cambios: nuevos desafíos de adaptación en Uruguay.* Montevideo, MGAP-FAO.

Bidegain, M., Crisci, C., del Puerto, L., Inda, H., Mazzeo, N., Taks, J. & Terra, R. (2012) *Clima de cambios: nuevos desafíos de adaptación en Uruguay.* Vol. I of *Variabilidad climática de importancia para el sector productivo.* Montevideo, MGAP-FAO.

Boerger, A. (1928) *Observaciones sobre agricultura. Quince años de trabajos fitotécnicos en el Uruguay.* Montevideo, Imprenta Nacional.

Cabrera, L. (2011) *Patrimonio y Arqueología en la región platense.* Montevideo, Universidad de la República.

CLAES/PNUMA/DINAMA. (2008) *Geo-Uruguay. Informe del estado del ambiente.* Montevideo, PNUD.

Conca, K. (2006) *Governing Water: Contentious Transnational Politics and Global Institution Building.* Cambridge, MIT Press.

Contreras, S. (2012) *Cuando la ciudad se convierte en río: una aproximación antropológica a las inundaciones en Mercedes – Uruguay.* PhD Thesis. Montevideo, Instituto de Ciencias Antropológicas.

Corena, C. (2018) *Gestión de agua en Ciudad del Plata. Identidad, representaciones, territorio.* Master's Thesis. Montevideo, Maestría en Antropología de la Cuenca del Plata, Facultad de Humanidades y Ciencias de la Educación, Universidad de la República.

CTMSG (Comisión Técnico Mixta de Salto Grande) (1977) *Aspectos de Impacto Ambiental. Ecosistemas afectados por la construcción de la represa de Salto Grande: Introducción a su Prospección Ecológica en Territorio Uruguayo.* Montevideo, Facultad de Humanidades y Ciencias, Universidad de la República.

Descola, P. (1996) Constructing Natures: Symbolic Ecology and Social Practice. In: Descola, P. & Pálsson, G. (eds.). *Nature and Society: Anthropological Perspectives.* London, Routledge.

De Torres Álvarez, M. F., Cruz, G. & Taks, J. (2007) Una aproximación a la comunicación social del clima en el caso del sistema pastoril del norte de Uruguay, Caso Salto. In:

Semana de Reflexión sobre cambio y variabilidad climática. Montevideo, Facultad de Agronomía, Universidad de la República.

De Torres Álvarez, M. F. & Piñeiro, D. (2013) Del cielo a la tierra: percepción ambiental de la ganadería. In: Picasso, V., Cruz, G. & Astigarraga, L. (eds.). *Cambio y variabilidad climática: respuestas interdisciplinarias*. Montevideo, Espacio Interdisciplinario, pp. 81–96.

Foladori, G. (2005a) El papel de la ciencia en la moderna conciencia ambiental – El caso de los desastres naturales. *Revista Saúde e Ambiente*. 6 (1): 51–57.

Foladori, G. (2005b) La enseñanza de Katrina. *Rebelión*. Available from www.rebelion. org/noticia.php?id=19789. Accessed December 7th 2018.

Foladori, G. (2005c) Una tipología del pensamiento ambientalista. In: Foladori, G. & Pierri, N. (eds.). *¿Sustentabilidad? Desacuerdos sobre el desarrollo sustentable*. Mexico, Miguel Angel Porrúa, Universidad Autónoma de Zacatecas, pp. 83–136.

Foladori, G. (1985) *Proletarios y campesinos*. Mexico, Universidad Veracruzana.

Foladori, G., Pierri, N. & Taks, J. (1996) *Metodología para el análisis de la percepción ambiental*. Working document num.20. Montevideo, Departamento de Sociología, Facultad de Ciencias Sociales, Universidad de la República.

Foladori, G., Pierri, N., Tommasino, H., Chang, M. & Taks, J. (2005b) Tres tesis básicas ocultas en la cuestión ambiental. In: Foladori, G. (ed.). *Por una sustentabilidad alternativa*. Montevideo, UITA, Universidad Autónoma de Zacatecas, pp. 52–59.

Foladori, G. & Taks, J. (2004) Um olhar antropológico sobre a questão ambiental. *Mana – Estudos de Antropologia Social*. 10 (2), pp. 323–348.

Foladori, G., Tommasino, H. & Taks, J. (2005a) La crisis ambiental contemporánea. In: Foladori, G. & Pierri, N. (eds.). *¿Sustentabilidad? Desacuerdos sobre el desarrollo sustentable*. Mexico, Miguel Angel Porrúa, Universidad Autónoma de Zacatecas, pp. 9–26.

García-Acosta, V. (2018) Anthropology of Disasters. In: Callan, H. (ed.). *The International Encyclopedia of Anthropology*. New York: John Wiley & Sons, Ltd., pp. 1622–1629.

García-Acosta, V. (2015) Historical Perspectives in Risk and Disaster Anthropology: Methodological Approaches. In: Wisner, B., Gaillard, J. C. & Kelman, I. (eds.). *Disaster Risk. Critical Concepts in the Environment*. London and New York, Routledge, vol. 3, pp. 271–283.

García Cartagena, M. & Taks, J. (2015) Transferencia internacional de políticas públicas y comunidades epistémicas: el caso del proyecto de Implementación de Medidas Piloto de Adaptación al Cambio Climático en las Áreas Costeras de Uruguay. In: Astigarraga, L., Terra, R., Cruz, G. & Picasso, V. (eds.). *Centro Interdisciplinario de Respuesta al Cambio y a la Variabilidad Climática: vínculos ciencia-política y ciencia-sociedad*. Montevideo, Espacio Interdisciplinario, Universidad de la República.

Giménez, A., Castaño, J. P., Baethgen, W. & Lanfranco, B. (2009) *Cambio climático en Uruguay, posibles impactos y medidas de adaptación en el sector agropecuario*. Serie Técnica N° 178. Montevideo, INIA.

Guigou, N. L. (2016) Antropología social en la nación uruguaya. In: *70 años. Facultad de Humanidades y Ciencias de la Educación*. Montevideo, Facultad de Humanidades y Educación, Universidad de la República.

Harvey, D. (2000) *Spaces of Hope*. Edinburgh, University Press.

Ingold, T. (2018) *Anthropology and/as Education*. London, Routledge.

Ingold, T. (2012) *Ambientes para la vida. Conversaciones sobre humanidad, conocimiento y antropología*. Montevideo, Trilce.

Ingold, T. (2000) *The Perception of the Environment*. London, Routledge.

Jacob, R. (1969) *Consecuencias sociales del alambramiento (1872–1880)*. Montevideo, Ediciones de la Banda oriental.

Kula, W. (1977) [1965] *Problemas y métodos de la historia económica*. Barcelona, Península.

Lettow, S. (2016) Turning the Turn: New Materialism, Historical Materialism and Critical Theory. *Thesis Eleven*. 140 (1), pp. 106–121.

Oliver-Smith, A. (2017) Adaptation, Vulnerability and Resilience: Contested Concepts in the Anthropology of Climate Change. In: Kopnina, H. & Shoreman-Ouimet, E. (eds.). *Routledge Handbook of Environmental Anthropology*. London, Routledge, pp. 206–218.

Oliver-Smith, A. & Hoffman, S. M. (2002) Introduction: Why Anthropologists Should Study Disasters. In: Hoffman, S. M. & Oliver-Smith, A. (eds.). *Catastrophe & Culture: The Anthropology of Disaster*. Santa Fe, Nuevo México, School of American Research, pp. 3–22.

ONU. (2015) *Informe de la evaluación interagencial sobre el estado de la reducción de riesgo de desastres en Uruguay*. Montevideo, Presidencia de la República, ONU.

OPP/BID/OEA. (1992) *Estudio ambiental nacional*. Washington, OEA.

Pi Hugarte, R. (1969) *El Uruguay indígena*. Montevideo, Nuestra Tierra.

Piperno, A., Quintans, F. & Conde, D. (coord.). (2015) Urban Waters in Uruguay: Progresses and Challenges to Integrated Management. In: *Urban Water Challenges in the Americas: A Perspective from the Academies of Sciences*. Mexico, IANAS & UNESCO, pp. 524–555.

Real de Azúa, C. (1984) *Uruguay ¿una sociedad amortiguadora?* Montevideo, CIEDUR/ EBO.

Renfrew, D. (2018) *Life Without Lead: Contamination, Crisis and Hope in Uruguay*. Oakland, University of California Press.

Ribeiro Coelho, H. (2003) *Las memorias de la memoria. El exilio de Darcy Ribeiro en Uruguay. Entrevistas*. Belo Horizonte, FALE/UFMG.

Romero Gorski, S. (ed.). (2014) *Dinámica cultural en la producción de salud y de riesgos*. Montevideo, Nordan.

Romero Gorski, S. (2010) Discusión conceptual antropológica en el marco del abordaje ecosistémico para el control del vector del dengue. In: Basso, C. (ed.). *Abordaje ecosistémico para prevenir y controlar al vector del dengue en Uruguay*. Montevideo, Universidad de la República.

Rossal, M. & Fraiman, R. (2007) Relaciones de intercambio en el tránsito urbano. *Anuario de Antropología Social y Cultural en Uruguay*. 6, pp. 155–162.

Rostagnol, S. (2016) Las antropologías de principios de siglo en Uruguay. In: *70 años. Facultad de Humanidades y Ciencias de la Educación*. Montevideo, Facultad de Humanidades y Educación, Universidad de la República.

Rostagnol, S. (2003) Representaciones y prácticas sobre sexualidad y métodos anticonceptivos entre hombres de sectores pobres urbanos. *Anuario de Antropología Social y Cultural en Uruguay*, pp. 39–55.

Taddei, R. (2017) *Meteorologistas e profetas da chuva: conhecimentos, práticas e políticas da atmosfera*. São Paulo, Terceiro Nome.

Taddei, R. (2012) The Politics of Uncertainty and the Fate of Forecasters. *Ethics, Policy & Environment*. 15 (2), pp. 252–267.

Taddei, R. (2005) *A comunicação social do clima. Esboço de uma sociologia do campo da comunicação meteorológica no Nordeste Brasileiro*. Paper presented at Simposio Internacional de Climatologia: A Hidro-meteorologia e os Impactos Ambientais em Regiões Semi-Aridas, Fortaleza, Brasil.

Taks, J. (2019) Transformaciones de la narrativa del cambio climático global en Uruguay. *Sociologías, Porto Alegre*. 21 (51), pp. 102–123.

Taks, J. (2001) Environment, Technology and Alienation: An Anthropological Study Among Modern Dairy Farmers in Uruguay. PhD Thesis. Manchester, University of Manchester.

Taks, J. (2000) Modernización de la producción lechera familiar y las percepciones del ambiente físico y social. *Anuario de Antropología Social y Cultural en Uruguay*. 1, pp. 109–125.

Thompson, D. (2018) Centralismo en percepción y respuestas a problemas ambientales en comunidades de Uruguay. In: Machado, C. (ed.). *Conflitos ambientais e urbanos. Pesquisas e Resistências no Brasil e Uruguai*. Rio Grande, FURG.

Thompson, D. (2016) Community Adaptations to Environmental Challenges Under Decentralized Governance in Southwestern Uruguay. *Journal of Rural Studies*. 43, pp. 71–82.

Thompson, D. & López Barrera, S. (2018) *Dolores After the Tornado: Collective Mobilization, Governance, and Adaptations in Southwestern Uruguay*. Paper presented at the Sixth Annual International Conference on Sustainable Development. Columbia, Columbia University.

Vidart, D. (1997) *Filosofía Ambiental. El ambiente como sistema*. Bogota, Editorial Nueva América.

Vidart, D. & Pi Hugarte, R. (1969) *El legado de los inmigrantes I y II*. Montevideo, Nuestra Tierra.

9   An epistemological proposal for the Anthropology of Disasters

The Venezuelan school[1]

Rogelio Altez

Introduction: the IDNDR effect

Like many other experiences, the Anthropology of Disasters in Venezuela was boosted by a catastrophe. Certainly, the landslides of 1999 that destroyed the central littoral (north of the country), represented a decisive opportunity for the development of research on the problem, both in applied and social sciences. Although the attention to the subject had existed for years, this event was a trigger for the study of disasters in Venezuelan Anthropology.

The 1999 disaster occurred in the closure of the International Decade for Natural Disaster Reduction (IDNDR),[2] the platform that stimulated institutions and researchers to specialize in the problem of risks. The UN decision came after evaluating the impact of disasters that previous decade, especially in affected countries with fewer resources. Precisely in Latin America, a series of disastrous events associated with natural phenomena produced great material losses and an elevated number of deaths.[3] Everything happened in a context of global recession and in an indebted region; each disaster directly affected the GDP, generating greater indebtedness.

The IDNDR produced supranational financing and imposed an agenda for that issue. In 1999, the International Strategy for Disaster Reduction, ISDR, was created, from which the United Nations Office for Disaster Risk Reduction was born. The ISDR implemented resources to establish international links and agreements, such as the World Conference on Disaster Reduction, held in Kobe in 2005, where the Hyogo Framework of Action, 2005–2015, was adopted. Thus, the institutions dedicated to the study of potentially destructive natural phenomena relied on direct and preferential attention from the states, and on an international platform that connected them to each other and with the formation of researchers. The Venezuelan Foundation for Seismological Research (FUNVISIS) was among the prestigious Latin American institutions that were directly stimulated by the UN agenda, and it will play an important role in the development of the subject in social sciences.[4]

The role of Latin America was essential on the discursive transformation of the multiple problem presented by disasters. Latin American researchers proposed the most important statement in the interpretative turn of the problem: "disasters are not natural", but are the catastrophic product of the intersection in time and space of one or several hazards with a vulnerable context (Maskrey, 1993; García-Acosta, 1996). This approach led to understand the study of hazards and vulnerability as a substantial element towards prevention. Technical expertise was associated with the comprehension of social and historical variables and, in that sense, the institutions dedicated themselves to the research of natural phenomena in long-term processes in order to estimate impacts on the return of these phenomena. In Venezuela, seismology set the tone.

In May 1997, the first Historical Seismology Workshop was held in Trujillo, at the Universidad de Los Andes (University of Los Andes).[5] The sessions revealed the importance of historical research of earthquakes for physical and social knowledge of its effects. Among the most important consequences of those workshops was the publication of three seismological catalogs when, up to that date, there was only one (Grases et al., 1999; Altez & Rodríguez, 2009). Since 1997, seismological cataloging has been a specialty in Venezuela, and the investigations carried out in this regard expanded the known seismic history.[6]

Thanks to this impulse, approaches, concepts, and methodologies on the study of risks were transformed, and new instances were founded. In 1992, the Foundation for Seismic Risk Reduction was created at the University of los Andes, with headquarters at the Geophysics Laboratory. In 1996, the Commission for Risk Mitigation was founded at the Universidad Central de Venezuela (Central University of Venezuela, UCV). In the Universidad Pedagógica Experimental Libertador (Libertador Experimental Pedagogic University) developed the National Project for Research, Education, and Risk Management in 2002. In the Universidad del Zulia (Zulia University) the research line Historical and Social Study of Disasters was assigned to the Center for Historical Studies and the Laboratory of History of Architecture since 2005 (Altez & Barrientos, 2008).

In Mérida, in 2007, the Research Center for Integral Risk Management was created and dedicated to multidisciplinary projects on disaster risk management, adaptation to climate change, and local sustainability. In 2010, at the Environmental Studies Center of the UCV, the course on Professional Improvement in the Reduction of Socionatural and Technological Risks in Environmental Management was created. All these initiatives have contributed knowledge, methodologies, and, above all, researchers to the field of disaster prevention and the transversal study of their problems.

In this context, we created the Anthropology of Disasters course at the School of Anthropology of the UCV in 2009: "The first university to offer a course on disasters anthropology was in Latin America", as García-Acosta said (2018: p. 6). To understand its place within the discipline in Venezuela it becomes necessary to go back in time and then return to the present. This chapter aims to explain the meaning of that subject, its results for this school, and especially its epistemological proposal, based on a critical distancing with functionalist empiricism (Godelier, 1976) and while being closer to a historical materialism without ideologies (Altez, 2016a).

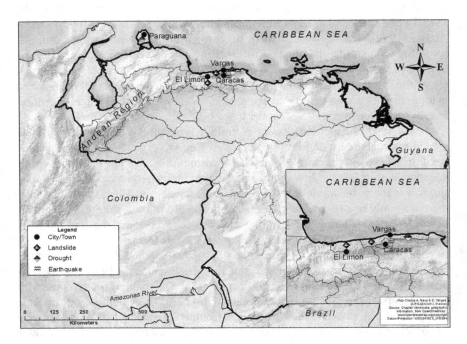

Map 9.1 Map Venezuela. Case studies and main areas mentioned

Anthropology becomes a school

The first ethnographers in Venezuela approached the indigenous realities according to the epistemological context of the 19th century. Positivists and evolutionists left us their thoughts on ethnology and "folklore" in a reality split among "caudillos" and liberalism. Gaspar Marcano (1850–1910), Lisandro Alvarado (1858–1929), Arístides Rojas (1826–1894), Tulio Febres Cordero (1860–1938), and Julio C. Salas (1870–1933) rode between costumbrism and science. Along with the German Adolfo Ernst (1832–1899), who was maybe the most qualified, they made great efforts in understanding the reality from an anthropological glance.

Only Rojas, Febres Cordero, and Ernst incorporated the observation of natural phenomena to their most passionate activities. We owe to that impulse of totalizing reality their chronologies and articles on earthquakes, climate, or geology, subjects that were mixed with reasoning about colonial society or the pre-Columbian past. They were polyhedral thinkers, without a unique field of study. However, their work vanished in time due to the absence of continuity, because the Venezuelan anthropology did not have an uninterrupted line from the 19th century to the present. The transversal view of phenomena ended up engulfed in the technical specialization, finally capitalized by the natural sciences.

Without a formal academic space, Venezuelan anthropologists developed scarce activity during the first half of the 20th century. Those who continued the ethnographic path dedicated their work to indigenous communities. A few interested in paleontology would exchange experiences with geologists, using archeological techniques and methods. Nevertheless, natural phenomena did not find space among their interests. They cannot be blamed: the phenomena and its effects were never a notable subject of the discipline.

Towards 1952–1953, in the boom of US oil investment in the country, the Department of Sociology and Anthropology was founded in the UCV, afterwards a School since 1957. Amidst the growing influence of the United States in the region, the anthropology of the north crossed glances with the institutionalization of the discipline in Venezuela. The emblematic publication of the Smithsonian Institution edited by Julian Steward (1940–1947), where he noted the absence of large studies on the ethnic groups of this territory (Torrealba, 1997), stimulated the development of an indigenist vocation still persists. Along with the professionalization of anthropology, the contemporary ethnology in Venezuela was also born (Margolies & Suárez, 1978), consolidating the descriptive archeology begun by the influential work of Cruxent and Rouse (1958, 1963).

The – almost global – university whirlwind of the 1960s found Venezuelan anthropology at the height of the Cold War. Descriptive archeology was disregarded, and Marxist Social Archeology burst with force (Vargas, 1998). In the same way, populist indigenism flourished. The anthropological practice, in any case, demonstrated a subsidiary character in relation to foreign theories (Torrealba, 1984). The professional pressure from the anthropologists ended up forcing their separation from the School of Sociology, and in 1986 it started its own life with four emblematic departments: Ethnology and Social Anthropology, Linguistics, Archaeology, and Physical Anthropology. All of them focused on traditional subjects of study: indigenous communities, descendants of enslaved, pre-Columbian societies, aboriginal languages, and forensic areas, among others. Fieldwork has been – and still is – the methodological identity of the discipline for most Venezuelan anthropologists. In none of these areas were natural phenomena involved as an analytical interest, beyond – of course – being treated as circumstantial aspects concomitant to archaeological objects of study.

In many ways, the Cold War and it ideological effects left a mark on the Venezuelan society. Academically speaking, its most distinctive trait can be found in the influence Marxism had on the development of social sciences. By way of example, the School of Sociology promoted the creation of the Department of Socio Historical Analysis, where a subject by the name of "Venezuelan Social Formation" was introduced from a materialistic point of view. Its name is eloquent. This subject, which is in the School of Sociology, is also shared between both schools up to the present, preparing anthropologists and sociologists in that perspective.

Other subjects where decisive as well for the education of anthropologists with a materialistic base, such as "Political Anthropology" or "Economic Anthropology", where the reading of Eric R. Wolf, Georges Balandier, Maurice Godelier,

Marvin Harris, Claude Meillasoux, Ángel Palerm and Josep Llobera, among others, complement the look already offered by the classical texts of Marx, Engels, and Gramsci, or the Frankfurt School. However, it is necessary to make an epistemological precision that is not always warned: Marxism is an ideology and historical materialism is an interpretative proposal; thus, are not the same. As an ideological perspective, Marxism has been highly significant in social sciences, and Venezuelan anthropology has not been immune to such influence. Latin American social archaeology represents, indeed, one of the most conspicuous examples; its ideological commitment has led them to political militancy and their arguments confirm it literally, with greater emphasis in the 21st century (see Sanoja & Vargas, 2004; Vargas, 2007).[7]

This School of Anthropology, of course, has certain peculiarities in comparison with other centers of anthropological training in the Western world. On one hand, it is a "school", and not a department or a research center. This marks a substantial difference, especially when compared to the teaching of anthropology in many European universities, where each one of the areas that constitute our school represent independent departments there. The place that each education center occupies in relation to the distribution of knowledge at each academic instance (school, research center, or department), also reveals the different conceptions of anthropology and where it is anchored epistemologically, according to the criteria that conceives it as a career or field of study.[8]

Our school developed its own way to teaching the discipline, focused only in its profession profile, without proposing to assist with its knowledge the rest of the faculty to which it is ascribed. It conceives anthropology as an independent and exclusive field, and under that conviction it was separated from the School of Sociology. It constituted the formation of anthropologists assuming the area of each department as indivisible parts of the discipline, and not as different specialties. Its success or departure from the original plan can be discussed; however, that was its most ambitious goal.

This possesses a peculiar interest for the epistemological problems of the discipline, perhaps as a legacy of its connection with sociology for several decades. Moreover, in its current conformation there are seven subjects that are common to both schools, taught in sociology until 1998. This marked interest in the theories and their most significant contents has been directly inherited and is a clear concern for our proposal on the Anthropology of Disasters.

The teaching of the discipline, without a doubt, is a heterogeneous reality that leads to equally heterogeneous results in its understanding and application. Anthropology is not a megalithic science, but a way of understanding social, historical, and symbolic processes from a broad theoretical, methodological, and interpretative basis. From this we can understand the differences between approaches, authors, currents, perspectives, techniques, concepts, theories, methods, and many other aspects that, instead of being common to all educational centers, are substantially different. Disasters will arrive at the school and to its peculiarities by the end of the 20th century.

A biographical matter

At the beginning of 1996, and thanks to the experience in documentary research, FUNVISIS commissioned us to investigate the Venezuelan 1812 earthquakes, the more devastating in the history of the country.[9] The approach to the context of 1812, the most critical moment of the independence process, allowed the development of a series of studies that combined the use of new and complementary interpretative tools for the historiography on that period. What we started then cultivated us in the field of historical seismology, the "semiology of earthquakes", as Guidoboni would say (1997).

This experience was, certainly, a transdisciplinary opportunity. This process initiated in 1996 in the midst of the IDNDR, the same year in which the first volume of *Historia y desastres en América Latina* (*History and disasters in Latin America*) was published, a book coordinated by Virginia García-Acosta that contributed to place our work within a specific line of research: the historical and social study of disasters.

Reading this work, it was possible to focus our research about 1812 as an historical analysis of a disaster, rebuilding its context to understand as the results of social and economic processes, where disasters constitutes a common thread (García-Acosta, 1996: p. 20). The extensive documentary investigation contributed on the one hand to seismology by concluding that on the afternoon of 26 March 1812, there happened not one, but two and even three earthquakes almost simultaneously (Altez, 2006; Choy et al., 2010). On the other hand, the investigation allowed us to define this conjuncture within the independence process as "the disaster of 1812" (Altez, 2006), and in this way to contribute to a new historiographical debate about the period.[10]

The transversal comprehension of disasters did not develop solely from our academic and professional training. Direct experience was equally decisive. In 1999, we experienced it when the whole region was devastated by one of the most important disasters in the history of Venezuela: "The Tragedy", as it is known by Venezuelan society since its impact by December of that year in the state of Vargas.

Up to 240 km² of affectation, 25% of the population displaced, US$ 2,069 million in losses, more than 20,000 homes damaged or destroyed, and the qualification as a "mass death disaster", represented its most evident impact (Altez, 2007). In the research activity, it produced effects as sensitive as in our own biography. The "Vargas case" ended up being an open school that allowed studies, theses, field works, and several specialized publications.

About this experience there was developed the proposal about an "unavoidable ethnography" (Altez & Revet, 2005; Altez, 2010; Altez & Osuna, 2018), a methodological resource that attempts to turn daily life into an object of study from an inevitable approach to everydayness from an anthropological perspective. Sandrine Revet proposed the *ethnologie des catastrophes* (Revet, 2008, 2013; Langumier & Revet, 2011; Langumier, 2008), an interpretative relationship that starts from the problem "d'ethnographier un événement auquel on n'a pas participé" (Revet, 2008: p. 3). Revet's research moreover contributed decisively to the founding of the anthropological studies about disasters in France. There is also the study by Paula Vásquez (2009) about the 1999 disaster; although far from the Anthropology of Disasters, it is a critical analysis of the event focused on power relations. The effects of this catastrophe, whose impact directly benefited researchers from all sciences and from different countries which provide a large amount of funding at all levels (institutional, national, international, and supranational), have reached unsuspected academic projections.

Regarding the long-range studies of this problem, at the School of Anthropology, we directed the theses of Klein (2007) and Vázquez (2011), and also accompanied Revet from the beginning of her formation as an anthropologist (Revet, 2004, 2006), till the development of her doctoral thesis (Revet, 2007). As an inhabitant of the area, we lived there from 1980 until 2011. As a survivor, the 1999 disaster was a biographical disruption that represented a critical impulse in the consolidation of the transversal studies on disasters.

Disaster arrives at the school

The foundation of the subject Anthropology of Disasters does not mean that it has been accepted since its first proposal. In fact, in 2003 we suggested it in the same department but it was rejected because it was understood that it did not represent a line of research in the discipline. Perhaps the advances of the discourse on disasters in the social sciences allowed a change of opinion on this matter.

On the first stage (2009–2011) four bachelor theses were produced. In its second stage (2015–2019) nine more theses were initiated; three were recently finished and the rest are expected to conclude soon. The works already provided a range from case studies (Padilla, 2011; Rodríguez, 2011), processes (Klein, 2007; Noria, 2011), to collective memory (Vázquez, 2011). Some of these works have reached bibliographic editions (Padilla, 2012; Rodríguez, 2018). Within the studies on development, there are investigations on anthropic disasters, the production of vulnerability, and disasters of mass deaths.

These investigations represent significant contributions to anthropological interpretations of disasters, not only having Venezuela as case studies. The work by María Victoria Padilla (2011, 2012) demystifies the disaster and famine that occurred in Paraguaná, in western Venezuela, between 1911 and 1912, an event that had never been formally investigated before. Emma Klein (2007, 2009) proposes conceptual contributions: the "distorted perception of risk", a category that calls for a debate and becomes useful in the critical perspective of risk production analysis. María N. Rodríguez (2011, 2018) conducted an in-depth investigation about the plague of locusts in Venezuela during the last decades of the 19th century, analyzing the material and social context of the disaster. This work is a transversal study over an unattended process in Venezuelan historiography.

Andrea Noria (2011) developed an investigation on the transformations of thought between colonial and republican society in Venezuela. Her study analyzes the interpretation of earthquakes among the scientists of the context, noticing the formalization of that thought in the institutionalization of sciences. It is an archeology of seismological thought carried out with the analytical tools of philosophical anthropology.

More recent theses reveal a preoccupation toward social problems strictly under the influence of the current context in Venezuela. These include the works by Diana Osuna (2019) about the mass deaths that occurred during the *Caracazo*,[11] and the study by Pedro Abreu (2019) on the disaster of shortages in Venezuela between 2000 and 2017.

On the other hand, it has also been possible to consolidate institutional relations thanks to the consolidation of the topic at the School of Anthropology. With CIESAS (Centro de Investigaciones y Estudios Superiores en Antropología Social, Center for Research and Advanced Studies in Social Anthropology), we founded two international research networks in 2015 and 2016.[12] With the Universidad de Sevilla (Sevilla University) a Permanent Seminar on Historic and Social Studies about Nature and Environment was created in 2011.

A long hiatus may be observed between the ethnographers of the 19th century and the arrival of the studies of disasters as a discipline in Venezuela. However, the proposal of transversal interpretation of the processes leading to catastrophes does not focus solely on phenomena. Other processes exclusively human also produce disasters, so the analytical focus is on historical, social, material, and symbolic processes, regardless of whether the phenomena intervene in them.

As can be seen, the Anthropology of Disasters has not been a subject created in isolation. Its process also shapes the biography of the discipline and our school, as much as our own biography. They are biographies that intertwine and determine each other. The establishment of this field of study has shown evolutions in its consolidation process as a field of anthropological research in Venezuela. Part of that process evinces the production of an approach based on materialism, but in a sustained dialog with other currents, and not necessarily in a belligerent way. In this evolution the discourse was conceptually transformed, and from there on a theoretical revision of the interpretative perspectives in this field of study was proposed as a space to share experiences and reasoning.

Paths to a materialistic approach

The epistemological precision that lies in the origin of our research about disasters begun with García-Acosta (1996). It is important to notice this because the path on the building our own theory is never independent from other theories or ways of interpreting. The process on this matter, the very sequence of that path, accounts for the epistemological resources that are used and chosen. To this effect, it should be mentioned that the education in anthropology and history, which had already provided a materialist interpretive platform, guarantees the understanding of disasters as processes. However, the fact that it became an object of study scarcely handled in the discipline, initially posed some methodological difficulties. The introduction by García-Acosta (1996) of the historical study of disasters placed the research on an incipient but solid field of study.

We are not Marxists, and we are neither orthodox nor excluders; therefore, work is not done from ideology, but from research. This starts from the possibility of reconstructing theories and advancing on the hermeneutical sedimentation produced through the epistemological articulation between theories. This materialistic foundation never excluded other currents; it has also borrowed from the structuralism of Lévi-Strauss or the semiotic frameworks of Geertz; from Koselleck's historicity; the revolutions of Hobsbawm; or the *longue durée* of Braudel. We are not tributaries of any ideology. The construction of a critical view about historical processes can only be achieved from the analysis, which means that it is from a theoretical formation and not from political commitments. Historical materialism, as with any other interpretative proposal, must not be treated as an obtuse logic. It could help to understand it in this way if we assume that materialism has long ceased to be only what Marx wrote in the mid-19th century.

Theories evolve once researchers stop taking them as a mandate. They also evolve because they feed on new ideas and reconstitute theirs. This is certainly the result of a permanent exercise of reading and debate, and of dialog and discussion. On the basis of these premises in the Anthropology of Disasters – as in every field of research – there must be theoretical debate; stop meeting to know "what" is being done, and start to reasoning about "what we think we are doing" when we do anthropology.

For all these reasons, it is pertinent to explain the epistemological basis for the understanding of disasters and all their variables. Moreover, in this explanation of course, there is discussion with other perspectives. However, everything starts from an undoubted recognition: the Anthropology of Disasters is the result of an equally theoretical positioning whose most decisive promoter is Anthony Oliver-Smith. If the subject has an analytical space it is because of his research. Thanks to his proposal, anthropology has understood that disasters evinces "the intersection between nature and culture [which have a] multidimensional condition [and that] in themselves include the process and the event" because they are, in any case, the result of processes (Oliver-Smith, 2002). The epistemological basis of this proposal, as has been explained many times, is founded upon political ecology (García-Acosta, 2004; Sökefeld, 2012; Díaz-Crovetto, 2015; Faas & Barrios, 2015; Faas, 2016; Baez Ullberg, 2017).

Along with Susanna Hoffman, Oliver-Smith has provided to the subject with the necessary continuity to make it a field of study of its own. Other colleagues, such as William Torry, Herman Konrad, or Mary Douglas, did not have the same intention, despite their influential works. However, the notable transversality of disasters has produced an interpretative field and a discourse that is equally transversal, and not exclusively anthropological (Faas, 2016). Conceptually, disasters and their variables have become more complex as a problem from diverse discursive contributions, as has happened with the case of La Red de Estudios Sociales en Prevención de Desastres en América Latina (Social Studies Network for Disaster Prevention in Latin America, LA RED), founded in Costa Rica in 1992.

LA RED was constituted by professionals from different areas of knowledge. Together with Oliver-Smith and García-Acosta, there are geographers, lawyers, philosophers, and engineers, all concerned and dedicated to the problem of disaster risk as researchers or consultants. Their original proposals transformed the interpretation and action on the subject, both in the academic space and on the institutional sphere. Perhaps that is why a conceptual mixture that does not clearly show any epistemological unity can be detected. However, it was through this effort that the entire world began to understand, finally, that disasters are not natural (Maskrey, 1993).

A great functionalist presence may be identified in this discursive mixture (Altez, 2009, 2016a), especially in relation to its most conspicuous conceptual proposals: risk and vulnerability. We distance critically from those proposals, so far as the process of evolution of an interpretative base has progressively transformed the discourse. That is why our work, for example, from the first moment used the term "social construction" at the same time as "production" (see Altez, 2006, 2009, 2010), searching to settle the perspectives with this object of polyhedral and multidimensional study. Social production and social construction, of course, are not synonyms, and it has been a methodological misuse to use them in that way. The difference between both terms is essentially epistemological, which leads analytically to conform different ways of interpretation of the processes that try to understand both concepts.

In any case, some premises were always clear to our point of view: we never shared the notion of "vulnerability angles" (Wilches-Chaux, 1993, 1998), a skillful methodological resource that, in its comfortable fragmentation of reality, does not help to understand it critically. Nor do we assume the common synonymy between vulnerability and poverty, which understands vulnerability as a diagnosis of reality (Altez, 2016b). As it demonstrated in Vargas 1999, disasters do not shock only one part of society: they impact all of it, and its effects are heterogeneous as the society itself is. What disasters show after its occurrence is the specific inequality within a specific society, which is not a universal reality. This is why poverty is not always the most affected part or the less resilient one; disasters impact without distinction of class, and the Vargas case is a good example of it (Altez, 2010).

Our approach found stability with the doctoral research and from the development of a long-term work over an extended period, which allowed us to approach in detail a deep analysis of vulnerability within a specific society: the colonial one in Venezuela (Altez, 2016a). Thanks to this investigation, a particular perspective on disasters analysis has been consolidated in our work.

Concepts and essential analytical categories

The materialistic approach in the Anthropology of Disasters can be explained from the following concepts, which are decisive to understanding disasters and their underlying variables – risk, vulnerability, hazards – from our perspective.

Production

The epistemological starting point is founded on the synonymy assumed between "*society, history, and existence*".[13] Following Godelier (1989) human beings are social animals, but unlike other species they "produce the society" and do not rely on it solely as a form of survival and biological reproduction. Production of society, as a human condition, supposes a different thing from associating by pure instinct or gonadal impulses. In the production of society through time the different forms of social organization of humanity are observed. Had these forms not transformed throughout their existence then humans would be like the rest of the social animals: they would have survived over the centuries with the same association structure and would have only developed some difference in relation to the distinct environmental constrictions, according to the place of settlement.

The production of human societies is the crystallization of different forms of solving the problem of survival and the domination of nature. As these do not happen in the same way in every society, nor has it been invariable in time, they do not operate through biological mechanisms. If so, they would have been transformed by adaptation, and not by social conflicts as in fact it has happened. The successive transformations of the different societies through time are – therefore – socially produced, and not biologically driven. Therein lies the difference between social animals with other species. The production of society and its transformations over time are the very history of humanity.

Materiality and totality

In the different ways of solving the problems imposed by nature before societies, we observe "materiality" (Godelier, 1989: pp. 20–22), All while understanding that the nature external to human beings operates on them as much as it does the other way around. This relationship, moreover, has not been unique over time nor between each society. The material production of our species is existence itself: the particular form to solve the problems imposed by nature, as well as the equally particular form deployed to transform nature in benefit of the different forms of social organization produced by humanity.

Those particular forms of social production of materiality are "the first premise of all human existence and, therefore, of all history" (Marx & Engels, 1974: p. 48). History is the human way to pass through time, and its existence, that social production which in turn produces society itself, is part of our nature. Existence is, therefore, history itself. To understand human existence means to comprehend history and, with it, society.

Humans produce society, materiality, and the relationship with nature. On doing so, they operate simultaneously on concrete and symbolic plans. "To the extent that man through this movement acts on the nature outside him and changes it, at the same time he changes his own nature" (Marx in Schmidt, 2014: p. 78). It is observed there the epistemological principle that gives meaning to the notion of "totality", a different interpretative sense from that of approaching reality by parts, or understanding that reality through a methodological segmentation of the existing relations between species and nature. The species is indivisible from nature, since it shapes it. The understanding of its relationship with nature, which is physically different to it, underlies the analysis of the historical production of materiality.

Human production results from "socially organized humankind in a double sense-active in changing nature, and in creating and re-creating the social ties that the effect the transformation of the environment" (Wolf, 1982: p. 74). The social organization deployed to face nature is historically created and re-created to transform the natural environment; thus, changes in the ways of transforming nature produces changes in social organization.

Within these essential premises, we found coincidences and distances with political ecology in the Anthropology of Disasters. According to Oliver-Smith (2002: p. 24), "Disaster occur[s] at the intersection of nature and culture and illustrate, often dramatically, the mutuality of each in the constitution of the other". This "intersection" coincides with Godelier's concept of materiality. For the French anthropologist, material reality is found in the border between nature and culture (1989: p. 21). Therefore, it could be said that disasters take place in materiality, or they occur because of problems, inefficiencies or waste in these materialities. Understanding of this lead, of course, to the analysis of historical and social processes underlying such problems.

However, political ecology is only focused on disasters associated with natural phenomena, preferably, or those that result from technological hazards (Hoffman & Oliver-Smith, 2002: pp. 4–5; Oliver-Smith, 2002: p. 25). Society, with or without "intentionality" (Hoffman & Oliver-Smith, 2002: p. 4), can become its own threat and trigger abrupt or slow disastrous processes, leading to serious losses of all kinds. Here we find the concept of "production" in all senses, and not only as something which operates in the intersection of nature and culture. Human society is in itself a product, and its conflicts and problems also evince processes that may crystallize in disasters.

Social construction

The concept of production, on the other hand, differs substantially from the concept of social construction proposed by García-Acosta (2005), which takes up, outlines, and extends with methodological clarity what was originally proposed by Mary Douglas, Niklas Luhmann, or Denis Duclos. It is not a simple semantic matter; the difference between production and construction is epistemological. Its proximity as words only match etymologically, not theoretically. Social construction of risk and cultural perception of risk, as theoretical proposal, go hand in hand, as García-Acosta explains, but in social construction underlies "the base itself that explains the disaster processes" (2005: p. 23).

As Mary Douglas said (2003: p. 38), "It seems that social construction and consensus greatly influence human perceptions". So, it is inferred that the specific forms of consciousness, of coding, and of "interpreting the event" are essential to understand the contextual relation with the disaster, that are in turn a decisive aspect in the social explanation that is given to it: "Whether disaster ensues depends to a large extent on how the event is interpreted". In this approach to the problem more attention is paid to the cognitive field, to the specific relationship between a society and an event or process. The concept of production, meanwhile, became an interpretive abstraction rather than an aspect of the process: it is the process itself.

Following Douglas, perception of risk is a social construction; following LA RED, material construction of risks is, as well, a social construction. Both approaches are complementary and come from different theoretical bases. In disaster's investigation, the notion of construction uses history as a resource, the ambitus where the construction takes place. Production, on the other side, is a concept indivisible from history because they are synonymous: history of human societies is the production itself of the species' existence. Here lies an epistemological difference; in the materialistic approach, it is not a methodological resource, it rather becomes part of an indivisible unity interpreted in that sense and understood that way.

There is no doubt about the usefulness of social construction as a key concept to understand problems about risks, vulnerability, and hazards. However, the concept, which could be understood as a "metaphor", like Hacking said, "once had excellent shock value, but now it has become tired" (1999: p. 35).[14] From a materialistic point of view, production is not a metaphor: it is a category with conceptual and methodological functions, and contains a double sense, as Marx proposed and Wolf explains: "active in changing nature, and in creating and re-creating the social ties that effect the transformation of the environment" (Wolf, 1982: p. 75).

It seems, according to the materialistic approach, that the notion of social construction could not be enough to explain long processes in human history, as production does. To understand analytically how a society produces and reproduces conditions that underlie vulnerability, materialistic theory could work better. Social construction, as an epistemological resource, offers limitations when facing long-term studies where the object to be interpreted has to be, essentially, how the existence of that society produces and reproduces its conditions of vulnerability.

Finally, where social construction sees a process (risk, vulnerability, hazards), it understands the verification of the construction of itself, the constructed object.[15] Production, on the other hand, is the nature of the social process, underlying history, the fate of the existence of society, and therefore is beyond crystallized forms of existence. It is also a continuum, the dynamics itself of the process.

Historicity and reproduction

The fundamental object in the Anthropology of Disasters is society, not the disaster itself. Certainly, a case study is as relevant as a long-term research; nevertheless, in both ways of analysis, societies must be regarded as an object, and not the specific fact that is observed. In this sense, the concept of "historicity" operates here with great methodological utility, and goes hand in hand with another essential category: "reproduction".

García-Acosta has raised it, even with the very concept of social construction: "The social construction of risks refers to the production and reproduction of vulnerability conditions" (García-Acosta, 2005: p. 23). The most characteristic object of study in the Anthropology of Disasters is centered there, and in our study of colonial society in Venezuela, it has been made clear (Altez, 2016a). Because of this, and in agreement with García-Acosta, it has been concluded that vulnerability is produced, but also reproduced, which can be noticed through the historicity of its most conspicuous indicators.[16]

The production of vulnerability is socially and materially determined, that is, it is historically determined. Therefore, we can observe its manifestations through time as indicators of its reproduction. However, these indicators (disasters, adversities, losses), although they reiterate certain aspects for centuries – the destruction of the same building by the impact of earthquakes, for example – do not mean the repetition of the problem, but the reproduction of the conditions that allow its manifestation. The historicity of the indicators should not be confused with the indicator itself. History does not repeat itself or repeat anything; and what is seen as facts associated with the same problem demonstrates its persistence and expounds its variability, as well as its transformation over time.

Repetition and reproduction do not mean the same, nor is the case with construction and production. In the reproduction of the conditions of vulnerability of a society we may also observe its transformation, since everything transforms, especially the human forms of existence. Such conditions of vulnerability, although they show some "durability" in history, contain in themselves their process of transformation in time. They run hand in hand with the transformation of society itself. In the words of Godelier (1976: p. 295), "the central problem of a science of history is to explain the appearance of the different social structures articulated in a determined and specific form, and the conditions of reproduction, transformation and disappearance of these structures and their articulation". As Wolf explains

> It is not the events of history we are after, but the process that underlie and shape such events. By doing so, we can visualize them in the stream of their development, unfolding from a time when they were absent or incipient, to when they become encompassing and general. We may then raise questions about proximate causation and contributory circumstances, as well as about the forces impelling the process toward culmination or decline.
>
> (Wolf, 1999: p. 8)

Critical windows

Disasters are "critical windows" that allow us to observe underlying processes. Looking through these windows societies are understood and events are explained. A disaster is not only what it is seen as the result of the confluence in time and space of hazards and vulnerable contexts; it is the empirical manifestation of a process, and that is why it is not enough to describe that manifestation: "how" the product came about must be analyzed, because its "why" is clear – the confluence mentioned previously. Hence, the analytical journey must lead to an understanding of the processes behind the event, and not only to the verification of the "why".[17]

Transformation

It may be observed in this analytical approach that the underlying process continues its movement. If a disaster manifests critically the production of conditions of vulnerability, by analytically traversing its manifestation, then the reproduction of those conditions can also be noted. For such conditions to disappear from history they must be perceived in the long-term nature of the event and corroborate that it is so. If a change has occurred in that society, it certainly does not happen because of the event, but rather through the dynamics of the process. That change is the most obvious form of "transformation".

Change must not be confused with transformation. Change is the crystallization of a process where the transformation occurs. This does not happen in generational speeds, or, as Geertz say (1995: p. 4): "apparently, [it] is not a parade that can be watched as it passes". Everything that changes in human processes is produced by transformation, and in the existence of societies this is historically produced. Changes in human societies are not substitutions, but expressions of structural transformations. Therefore, they cannot be assessed empirically or in fieldwork, but only through the analysis of long-term processes, or by deepening in critical junctures that reveal strong indicators of this process. We believe our latest study on the 1812 disaster in Venezuela (Altez, 2015) has contributed to this.

Phenomena and facts

Societies change because they are transformed, and this is manifested always through facts. The development of societies, their relations, representations, material, or objective conditions of existence, are "facts", never phenomena. A phenomenon simply "is", which means that it is something from itself.[18] That is why natural phenomena operate by causalities different to society, while everything that is socially human is also historical, that is to say: it does not exist simply by itself, but rather it is a historical and social product. In this epistemological difference it must be assumed that facts should not be confused with phenomena, and that social and historical facts will never be nor have they ever been phenomena. Human society can only produce facts, not phenomena. If phenomena are "the self-showing in itself", cannot be historically or socially produced: they can only be natural. That is why phenomena must always be "natural phenomena".

However, when natural phenomena intersect with societies they are incorporated into history; it is part of their process, or, together with society, form the process itself, because all this represents an indivisible unit. As Godelier says (1989: p. 21), this nature outside man is not outside of culture, society, or history. What the phenomenon drives in a context is a product of the historical, material, and social process which appear, and since its emergence is already a part of that historical process, thus becomes a fact. A phenomenon which does not interact with human society is only a phenomenon, and nothing more. When humans interact with phenomena, passively or actively, both are conforming a process and producing a greater one. Therefore, everything that is produced by the interaction with the phenomenon becomes a fact, as the phenomenon itself from that moment.

These facts, moreover, transcend their moment and shape history. In this way, in the material or subjective effects which are produced by the emergence of a potentially destructive phenomenon, the fact is found beyond its moment, of its instance, of the event. The concept of fact in here is articulated epistemologically with history, society and existence.

Relations and power relations

All social facts are "relations" as well. And every relation in human existence is social, therefore it has content. Such content, in turn, is contextually determined. This means that these relationships have symbolic, subjective, historical, and affective contents such as all kinds of human conditions that are historically and culturally produced. Hence, societies produce these relationships with nature as with other societies: between individuals, with the past, with the present, with time, and with the universe.

That is why a natural hazard, for example, is not an entity external to the society that suffers it, but has been shaped historically, materially, and symbolically as a hazard, as a feasible adversity. A hazard is the result of a relationship, and, like all human relations, it is contextually determined and its content is not invariable in history nor is its meaning universal. It is transformed like all human relationships, and, in this case, because it is a natural phenomenon, it is transformed symbolically and materially. The phenomenon may be the same, but its meaning is historically susceptible; therefore, it is not eternal. That is why hazards do not mean the same thing over time or culturally, and their condition of feasible adversity may also be transformed historically. Hence, some hazards that were once fearsome have now disappeared; while, on the other hand, societies have been able to produce new hazards that did not exist before.

A hazard is equally an abstraction that fulfills methodological functions, which means that it should not be understood as a "factor" whose nature is adversity. Nor do hazards derive exclusively from relationships produced within nature. Societies, no doubt, can stand as hazards either before other societies or even before themselves.

When a society objectively produces its interests in relation to other societies, it is able to satisfy them concretely. This can lead to many forms of exchanges, but also invasions, subjugations, exploitations, wars, and exterminations. Thus, society is in itself a hazard, not a natural one, but a hazard socially produced. On the other hand, a sector within a society can also erect itself as a hazard to the rest of society, producing crises and adversities of serious losses. A society transformed into a hazard, to other societies or to itself, is also the product of a historically and socially produced relationship. Everything which produces a hazard is directly proportional to the conditions of vulnerability. Therefore, there must always exist a relation between hazard and vulnerability.

Social relations are also "power relations", especially in class societies (Poulantzas, 1986); however, in that sense, it is a rule in every kind of human society: castes, lineages, highly stratified societies, bands, or any other. Power is a structure of social relations, which means it is a structure of human society. There is no society without power.

Following this argument, power relations and all its effects in society become an analytical course, essentially, in front of a case study, since power relations underlie the interests that produce materiality, for example, and with it the material conditions of vulnerability in a society (Altez, 2010). In modern societies and with proven evidence in the case of Latin America, for example, power relations have also deepened and exacerbated vulnerability conditions in every sense, especially in ideological and subjective levels, capitalizing these conditions basically for the benefit of the reproduction of political interests. Behind the analysis of these relationships will undoubtedly be found the most conspicuous causalities of disasters.

Finally, our analytical perspective is based on the epistemological articulation of the categories that have been presented here. Every one of them (materiality, totality, production, reproduction, historicity, transformation, facts, relations, and power relations), contained as well within the synonymy society-history-existence, leads to a critical understanding of the proper variables of disasters. Disasters, as we understand them, are critical windows that allow to observe underlying processes, social, and historical.

Behind a disaster also underlies its meaning. Following what became proposed by Lévi-Strauss (1987: p. 32), every apparent disorder has an underlying order; therefore, in the characteristic shudder of a disaster, in that disruptive disorder, its meaning can be found: "it is absolutely impossible to conceive the meaning without order" (p. 33). Every meaning is contextually determined, so that the meaning of a disaster must be understood as well in the context in which it takes place. The context, no doubt, gives sense to meaning. In correspondence with this, when it is proposed to understand a disaster, it will be developing a semiological analysis. Anthropologists go in search of that meaning, which occurs during its manifestation, as well as that which happens afterwards in memories or forgetfulness. In its contextual determination, when the sense of meanings is symbolically articulated, it is also historically, materially and socially conditioned.

The vulnerable context of the Anthropology of Disasters in Venezuela

Our approach has been based upon a reflection on a process that is still growing. Not only because the Anthropology of Disasters is a new field of study and still in consolidation, but because it is being conformed; it is not finished, but in process. However, this epistemological and academic process is threatened; Venezuelan society has produced itself as a vulnerable context and as its own hazard. We live in a disastrous conjuncture in full force, with effects that can be observed in the medium and long term.

The development of the field has not survived undamaged from the context of the country in recent years. The deterioration of universities and research is directly proportional to the economic and structural deterioration of society. Students trained in the subject have continued their profession in other countries or flee very quickly, or they have not been able to develop their career in any way. There are no possibilities for academic or material growth in a country without opportunities for university life.

The survival of the subject depends on isolated studies, without institutional funding, and sustained by personal efforts. Every day it becomes more difficult to publish due to the lack of resources and the disappearance of private publishers. Journals have been discontinued and libraries are dramatically unattended for years. Our connection with the outside academic world is reduced to individual achievements.

In spite of everything, our analytical approach continues growing; it does such in a reality that surrounds and presses it as a vulnerable context and object of study at the same time. The analytical deployment is a permanent exercise in this catastrophic daily life. We live practicing an "inevitable ethnography" applied to the understanding of processes that produce and reproduce vulnerability in front of our eyes.

However, the Anthropology of Disasters in Venezuela, its subject and its field of study, continues its journey. Everything is taking its place, since its inception and its consolidation in the School of Anthropology, in dialog and epistemological articulation with other currents and tendencies, such as that of Oliver-Smith, which has given rise to this field of study in the discipline. Within the theoretical debate lies the evolution of theories and methods, of reasoning and discourses in general. If it is assumed that we have a particular approach, then this primarily means an invitation for debate, and not a ditch that marks frontiers. We are the product of a similar invitation, the one that Virginia García-Acosta extended in 1996 when she presented to Latin America the historical study of disasters. It is thanks to the privilege of having read her work, as well as her unparalleled academic generosity, that we are here today.

Notes

1 The author would like to express his gratitude to Diana Osuna for her contribution in the translation of this work, and to Virginia García-Acosta, for the extraordinary editing work and her always-wise advice.
2 Resolution 44/236 of 1989, whereby the General Assembly of the United Nations proclaims the IDNDR as of 1 January 1990.
3 Between 1982 and 1983, the Andean region was severely affected, due to the ENSO phenomenon, with floods, landslides and droughts. In 1983 took place the earthquake that destroyed Popayán, and in 1985 the city of Armero, also in Colombia, was devastated by the eruption of the Nevado del Ruiz volcano, leaving more than 20,000 dead. That same year a big earthquake shook Mexico City and Chile suffered one of great magnitude as well. There were also disasters due to natural phenomena in 1986 in El Salvador, and Nicaragua and Costa Rica were affected by hurricane Joan in 1988; see Lavell (2005). To Lavell's summary, we can add the Gilberto hurricane, with serious

consequences in Central America in 1988, or the landslides in El Limón, Venezuela, in 1987. Gilberto alone, for example, caused losses up to U\$ 5 billion to the Central America countries.

4 Lavell (2005: p. 7) lists those institutions: "FUNVISIS in Venezuela, the Peruvian Institute of Geophysics and the Regional Center for Seismology for South America, CERESIS, in Peru; the Institute of Geosciences at the University of Panama, the School of Geology at the University of Costa Rica, nowadays the Central American School of Geology; the National Institute of Seismology, Volcanology and Meteorology in Guatemala, the Faculties of Engineering at the University of Costa Rica, the University of Chile, the National Autonomous University of Mexico and the National University of Engineering of Peru".

5 There are already five editions since then: 1997, 2000, 2002, 2009, and 2012.

6 Between the 1999 and 2009 catalogs, up to 398 new earthquakes were discovered in the 20th century.

7 Latin American Social Archeology is formed in the 1960s with a Marxist approach of critical manifestation to North American schools, dedicated to the reconstruction of social contexts of the observed past, either pre-Columbian or colonial. Its interpretations are close to the Latin American dependency theory, with scientific nuances. They take on the direct influence of European researchers, such as Gordon Childe and Andre Leroy-Gourham, as well as theorical linkage with Leslie White and Betty Meggers, or with André Gunder Frank. The most representative authors of this line are as follows: Venezuela, Mario Sanoja and Iraida Vargas; Peru, Luis Lumbreras; Chile-Mexico, Luis Felipe Bate; Dominican Republic, Marcio Veloz Maggiolo; and Mexico, Manuel Gándara.

8 Some examples in Ibero-American spaces: in the Universidad de Granada (Granada University), Physical Anthropology is taught in the Faculty of Medicine, and Social Anthropology in the Faculty of Philosophy and Literature; in the Universidad Complutense de Madrid (Complutense University of Madrid) the degree in Archeology belongs to the Faculty of Geography and History, while the degree in Social and Cultural Anthropology is in the Faculty of Legal and Political Sciences; in the Universidad de Los Andes (University of Los Andes) of Colombia, the Department of Anthropology is located in the Faculty of Social Sciences, as a Bachelor's Fegree; the Universidad de Chile (Chile University) has a Department of Anthropology in its Faculty of Social Sciences that offers specializations in Social and Physical Anthropology and Archeology; in the Universidad de Buenos Aires (University of Buenos Aires), the Department of Anthropological Sciences is located in the Faculty of Philosophy and Literature, where is possible to follow the sociocultural orientation or the archeological one. The closest program to that of the School of Anthropology of the UCV is offered by the degree in Anthropology of the Universidad Nacional Autónoma de México (National University of Mexico), anchored to the Faculty of Political and Social Sciences.

9 FUNVISIS was carrying out since 1995 a seismological investigation on the 1812 earthquake effects, and needed the expertise of a documentary researcher; this is the reason why they hired me. The project was titled *Neotectonic study and geology of active faults in the foothills of the southern Andes of Venezuela*, and thanks to this research it was possible to determine, as we will comment later, that on 26 March 1812, there was not one earthquake, but two, at least, a scientific result that changed the history of seismicity in the country. This research determined my training in the subject of disasters.

10 The relation with seismologists and geologists also led to a production of multidisciplinary works that contributed with transversal results over problems that – until then – had not been treated this way by Venezuelan scientists.

11 "Caracazo" is the name by which is known the social outbreak that took place in February and March of 1989 in Caracas and other cities of the country when a wave of looting represented the largest social manifestation in contemporary Venezuelan history against

the imposed economic measures by the Carlos Andrés Pérez government. The armed response to the protests ended the lives of hundreds of people in acts of generalized violence, extrajudicial executions, covert assassinations, and repression with firearms. Most of the corpses were buried in mass graves with no major formality, and the identification processes of the victims were flawed from the beginning until the exhumation of the remains located in those graves was decided under the Chávez government in 2009. Analytical attention to the case from the tools of the Anthropology of Disasters supposes, among other things, the application of the definition of "disaster of mass deaths" to the event, as well as the analytical and critical reconstruction of the process.

12 In 2015 we created the International Network of Seminars on Historical Studies on Disasters at CIESAS headquarters, Mexico City, with researchers from Spain (Armando Alberola, University of Alicante), Mexico (Isabel Campos Goenaga, National School of Anthropology and History; Luis Arrioja, El Colegio de Michoacán, Raymundo Padilla Lozoya, University of Colima); Chile (Andrea Noria, Autonomous University of Chile); and Venezuela (Rogelio Altez, UCV). Among its objectives was the development of a larger project dedicated to the analysis of risk, vulnerability, and disasters from a historical perspective. With this support, it was possible to create in 2016 the Thematic Network of Interdisciplinary Studies on Vulnerability, Social Construction of Risk and Natural and Biological Hazards, funded by the Consejo Nacional de Ciencia y Tecnología, CONACYT, of Mexico. This network achieved several collective publications (Arrioja & Alberola, 2016; Altez & Campos, 2018).

13 Facing possible epistemological confusions, it is convenient to specify the meaning which we observe in "production" as a category and its analytical influence on our approach. Marx took "social production" as a starting point for the analysis of "man" history (see: Marx, 1989: p. 6); thus, he observed the material production of existence itself as "making history" (Marx & Engels, 1974, I: p. 26). That is why Wolf explains the importance of the term when Marx indicates that he used it to designate "this complex set of mutually dependent relations among nature, work, social labor and social organization" (Wolf, 1982: p. 74). Continued Wolf: "The term *production* expressed for him [Marx] both this active engagement with nature and the concomitant 'reproduction' of social ties". The epistemological synonymy that we propose, "history-existence", contains itself the production and the reproduction of everything which is human, and for that reason we assume it as an interpretative articulation. Human production is unfailingly social, and therefore historical. When we said "human production", we also said "historical, social, symbolic, material production", and everything that results of the existence of our species: "to the contrary to other social animals, men are not happy with living in society, they produce society for living" (Godelier, 1989: p. 17). In that sense, we think that vulnerability, risk, and hazards are human products, social relations, and historical results, and are not disabilities associated to poverty or exclusively determined for inequalities.

14 "Construction has been trendy. So many types of analyses invoke social construction that quite distinct objectives get run together" (Hacking, 1999: p. 35).

15 "Process and product are both part of arguments about construction. The constructionist argues that the product is not inevitable by showing how it came into being (historical process), and noting the purely contingent historical determinants of that process" (Hacking, 1999: p. 39).

16 In this work we propose, for example, that when the same church is destroyed by regular manifestations of the same phenomenon for several centuries, certainly, history is not "repeating" itself, but the historicity of a condition is being demonstrated. The church becomes an indicator of a process in which conditions of vulnerability are produced and, above all, reproduced. This indicator becomes, therefore, the demonstration of the historicity of that vulnerability. The research, focusing on the current Venezuelan regions between the 16th and 19th century, provides documented information about the problem on all the major cities and regions.

17 For most researchers in Anthropology of Disasters, this process is in itself the social and material construction of risks; for our approach, in the historical and social process that underlies the causality of that result, it is not a deductible aspect of reality, it is the real logic behind the apparent (following Godelier, 1976).

18 Or, as Heidegger described, "the self-showing in itself" (2010: p. 31).

References

Abreu, P. (2019) *El desastre de la escasez. Análisis de la producción de vulnerabilidad en el contexto de desabastecimiento de Venezuela: 2000–2017*. BA Thesis, Caracas, Escuela de Antropología, Universidad Central de Venezuela.

Altez, R. (2016a) *Historia de la vulnerabilidad en Venezuela. Siglos XVI–XIX*. Madrid, CSIC-Universidad de Sevilla-Diputación de Sevilla.

Altez, R. (2016b) Aportes para un entramado categorial en formación: vulnerabilidad, riesgo, amenaza, contextos vulnerables, coyunturas desastrosas. In: Arrioja, L. & Alberola, A. (eds.). *Clima, desastres y convulsiones sociales en España e Hispanoamérica, siglos XVII–XX*. Zamora and Alicante, El Colegio de Michoacán, Universidad de Alicante, pp. 21–40.

Altez, R. (2015) *Desastre, independencia y transformación. Venezuela y la Primera República en 1812*. Castelló, Universidad Jaume I.

Altez, R. (2010) Más allá del desastre. Reproducción de la vulnerabilidad en el estado Vargas (Venezuela). *Cahiers des Amériques Latines*. 65, pp. 123–143.

Altez, R. (2009) Ciclos y sistemas versus procesos: aportes para una discusión con el enfoque funcionalista sobre el riesgo. *Desacatos Revista de Antropología Social*. 30, pp. 111–128.

Altez, R. (2007) Muertes bajo sospecha: investigación sobre el número de fallecidos en el desastre del estado Vargas, Venezuela, en 1999. *Cuadernos de Medicina Forense*. 13, pp. 255–268.

Altez, R. (2006) *El desastre de 1812 en Venezuela: sismos, vulnerabilidades y una patria no tan boba*. Caracas, Fundación Empresas Polar-Universidad Católica Andrés Bello.

Altez, R. & Barrientos, Y. (coord.). (2008) *Perspectivas Venezolanas sobre Riesgos: Reflexiones y Experiencias, Volumen 1*. Caracas, Universidad Pedagógica Experimental Libertador.

Altez, R. & Campos, I. (ed.). (2018) *Antropología, historia y vulnerabilidad. Miradas diversas desde América Latina*. Zamora, Mexico, El Colegio de Michoacán.

Altez, R. & Osuna, D. (2018) Vivir entre muertes masivas: sociedad y vulnerabilidad en Venezuela, 1999–2012. In: Altez, R. & Campos, I. (eds.). *Antropología, historia y vulnerabilidad. Miradas diversas desde América Latina*. Zamora, Mexico, El Colegio de Michoacán, pp. 193–228.

Altez, R. & Revet, S. (2005) Contar los muertos para contar la muerte: Discusión en torno al número de fallecidos en la tragedia de 1999 en el estado Vargas. *Revista Geográfica Venezolana*. Special Issue, pp. 21–43.

Altez, R. & Rodríguez, J. A. (coords.). (2009) *Catálogo Sismológico Venezolano del siglo XX. Documentado e ilustrado*, 2 vols. Caracas, Fundación Venezolana de Investigaciones Sismológicas.

Arrioja, L. & Alberola, A. (eds.). (2016) *Clima, desastres y convulsiones sociales en España e Hispanoamérica, siglos XVII–XX*. Zamora and Alicante, El Colegio de Michoacán, Universidad de Alicante.

Baez Ullberg, S. (2017) La Contribución de la Antropología al Estudio de Crisis y Desastres en América Latina. *Iberoamericana–Nordic Journal of Latin American and Caribbean Studies*. 46, pp. 1–5.

Choy, J. E., Palme, C., Guada, C., Morandi, M. & Klarica, S. (2010) Macroseismic Interpretation of the 1812 Earthquakes in Venezuela Using Intensity Uncertainties and A Priori Fault-Strike Information. *Bulletin of the Seismological Society of America.* 100, pp. 241–255.

Cruxent, J. M. & Rouse, I. (1963) *Arqueología de Venezuela.* Caracas, Instituto Venezolano de Investigaciones Científicas.

Cruxent, J. M. & Rouse, I. (1958) *An Archeological Chronology of Venezuela.* Washington, Panamerican Union (Science Monographs).

Díaz-Crovetto, G. (2015) Antropología y catástrofes: Intersecciones posibles a partir del caso Chaitén. *Justiçia do Direito.* 29, pp. 131–144.

Douglas, M. (2003) *Risk Acceptability According to the Social Sciences.* London, Routledge.

Faas, A. J. (2016) Disaster Vulnerability in Anthropological Perspective. *Annals of Anthropological Practice.* 40, pp. 14–27.

Faas, A. J. & Barrios, R. E. (2015) Applied Anthropology of Risk, Hazards and Disasters. *Human Organization.* 74 (4), pp. 287–295.

FUNVISIS. (1997) *Estudio neotectónico y geología de fallas activas en el piedemonte surandino de los Andes venezolanos.* Caracas, Fundación Venezolana de Investigaciones Sismológicas.

García-Acosta, V. (2018) Anthropology of Disasters. In: Callan, H. (ed.). *The International Encyclopedia of Anthropology.* New York, John Wiley & Sons, Ltd., pp. 1622–1629.

García-Acosta, V. (2005) El riesgo como construcción social y la construcción social del riesgo. *Desacatos. Revista de Antropología Social.* 19, pp. 11–24.

García-Acosta, V. (2004) La perspectiva histórica en la Antropología del riesgo y del desastre. Acercamientos metodológicos. *Relaciones. Estudios de historia y sociedad.* XXV (97), pp. 123–142.

García-Acosta, V. (1996) Introducción. In: García-Acosta, V. (coord.). *Historia y desastres en América Latina.* Bogotá, LA RED-CIESAS-Tercer Mundo Editores, vol. I, pp. 15–37.

Geertz, C. (1995) *After the Fact: Two Countries, Four Decades, One Anthropologist.* Cambridge, Harvard University Press.

Godelier, M. (1989) *Lo ideal y lo material.* Madrid, Taurus Humanidades.

Godelier, M. (1976) Antropología y economía. ¿Es posible una antropología económica? In: Godelier, M. (comp.). *Antropología y economía.* Barcelona, Anagrama, pp. 279–356.

Grases, J., Altez, R. & Lugo, M. (1999) *Catálogo de sismos sentidos o destructores de Venezuela, 1530–1998.* Caracas, Academia Nacional de Ciencias Físicas, Naturales y Matemáticas.

Guidoboni, E. (1997) Breve premessa sulla sismologia storica: una sismologia, una storia. In: Boschi, E., Guidoboni, E., Valensise, G. & Gasperini, P. (eds.). *Catalogo dei forti terremoti in Italia dal 461 a.C al 1990.* Roma, Istituto Nazionale di Geofisica, pp. 25–29.

Hacking, I. (1999) *The Social Construction of What?* Cambridge, Harvard University Press.

Heidegger, M. (2010) *Being and Time.* Albany, State University of New York Press.

Hoffman, S. M. & Oliver-Smith, A. (eds.). (2002) *Catastrophe & Culture: The Anthropology of Disaster.* Santa Fe, Nuevo Mexico, School of American Research.

Klein, E. (2009) Percepción distorsionada en la construcción social del riesgo. *Tierra Firme.* 107, pp. 219–228.

Klein, E. (2007) *Percepción sin Memoria y Vulnerabilidad Estructural en la Construcción Social del Riesgo: el caso general de las comunidades del estado Vargas venezolano.* BA Thesis. Caracas, Escuela de Antropología, Universidad Central de Venezuela.

Langumier, J. (2008) *Survivre à l'inondation: pour une ethnologie de la catastrophe.* Lyon, ENS Editions.

Langumier, J. & Revet, S. (2011) Une ethnographie des catastrophes est-elle posible? Coulées de boue et inondations au Venezuela et en France. *Cahiers d'anthropologie sociale.* 7, pp. 77–90.

Lavell, A. (2005) Los conceptos, estudios y la práctica en el tema de los riesgos y desastres en América latina: evolución y cambio, 1980–2004: El rol de LA RED, sus miembros e instituciones de apoyo. In: *La gobernabilidad en América Latina.* San Jose de Costa Rica, Secretaría General FLACSO, pp. 2–66. Available from http://biblioteca virtual. clacso.org.ar/ar/libros/flacso/secgen/Lavell.pdf. Accessed October 31st 2019.

Lévi-Strauss, C. (1987) *Mito y significado.* Madrid, Alianza Editorial.

Margolies, L. & Suárez, M. M. (1978) *Historia de la etnología contemporánea en Venezuela.* Caracas, Universidad Católica Andrés Bello.

Marx, K. (1989) *Contribución a la crítica de la economía política.* Moscow, Progreso.

Marx, K. & Engels, F. (1974) *Obras escogidas*, vol. I. Moscow, Progreso.

Maskrey, A. (comp.). (1993) *Los desastres no son naturales.* Bogota, LA RED-Tercer Mundo Editores.

Noria, A. (2011) *Miradas entre ruinas: La transformación en la interpretación sobre la naturaleza en la sociedad venezolana del siglo XIX a través del pensamiento sismológico.* BA Thesis. Caracas, Escuela de Antropología, Universidad Central de Venezuela.

Oliver-Smith, A. (2002) Theorizing Disasters: Nature, Power, and Culture. In: Hoffman, S. M. & Oliver-Smith, A. (eds.). *Catastrophe & Culture: The Anthropology of Disaster.* Santa Fe, Nuevo Mexico, School of American Research, pp. 23–47.

Osuna, D. (2019) *La ruta del Caracazo. Crítica de un desastre de muertes masivas.* BA Thesis. Caracas, Escuela de Antropología, Universidad Central de Venezuela.

Padilla, M. V. (2012) *El año del hambre La sequía y el desastre de 1912 en Paraguaná.* Mérida, Instituto de Cultura del Estado Falcón.

Padilla, M. V. (2011) *El año del hambre: La sequía y el desastre de 1912 en Paraguaná.* BA Thesis. Caracas, Escuela de Antropología, Universidad Central de Venezuela.

Poulantzas, N. (1986) *Poder político y clases sociales en el estado capitalista.* Mexico, Siglo XXI Editores.

Revet, S. (2013) "A Small World": Ethnography of a Natural Disaster Simulation in Lima, Peru. *Social Anthropology/Anthropologie Sociale.* 21, pp. 38–53.

Revet, S. (2008) L'ethnologue et la catastrophe. *Problèmes d'Amérique latine.* 69, pp. 1–24.

Revet, S. (2007) *Anthropologie d'une catastrophe. Les coulées de boue de 1999 au Venezuela.* Paris, Presses Sorbonne Nouvelle.

Revet, S. (2006) Le risque négocié. Conflits et ajustements autour de la reconstruction de Vargas (Venezuela). *Autrepart.* 37, pp. 163–181.

Revet, S. (2004) ¿Quién soy? ¿Quiénes somos? Entre categorización y estigma. ¿Cómo gestionan sus identidades los venezolanos damnificados? *Revista Venezolana de Economía y Ciencias Sociales.* 10, pp. 39–57.

Rodríguez, M. N. (2018) *Plagas, vulnerabilidades y desastres agrícolas: La sociedad venezolana a fines del siglo XIX.* Caracas, Academia Nacional de la Historia.

Rodríguez, M. N. (2011) *De plagas, desastres agrícolas y vulnerabilidades: Una mirada antropológica a la plaga de langostas que invadió el territorio venezolano entre 1883 y 1890.* BA Thesis. Caracas, Escuela de Antropología, Universidad Central de Venezuela.

Sanoja, M. & Vargas, I. (2004) *Razones para una revolución.* Caracas, Monte Avila Editores.

Schmidt, A. (2014) *The Concept of Nature in Marx.* London, Verso Books.

Sökefeld, M. (2012) Exploring the Link Between Natural Disasters and Politics: Case Studies from Peru and Pakistan. *Scrutiny. A Journal of International and Pakistan Studies.* 5, pp. 71–101.

Steward, J. H. (1940–1947) *Handbook of South American Indians*. Washington, Smithsonian Institution.

Torrealba, R. (1997) Antropología. *Diccionario de Historia de Venezuela*. Caracas, Fundación Polar. 1, pp. 167–169.

Torrealba, R. (1984) Los marcos sociales e institucionales del desarrollo científico en Venezuela: El caso de la antropología social. In: Vessuri, H. M. (ed.). *Ciencia académica en la Venezuela moderna. Historia reciente y perspectivas de las disciplinas científicas*. Caracas, Fondo Editorial Acta Científica Venezolana, pp. 213–235.

Vargas, I. (2007) Antropólogos y antropólogas para qué. In: Meneses, L., Gordones, G. & Clarac, J. (eds.). *Lecturas antropológicas de Venezuela*. Merida, Editorial Venezolana, pp. 17–32.

Vargas, I. (1998) La profesionalización de la arqueología en Venezuela, 1950–1995. In: Amodio, E. (ed.). *Historias de la antropología en Venezuela*. Maracaibo, Universidad del Zulia, pp. 345–354.

Vásquez, P. (2009) *Poder y catástrofe. Venezuela bajo la tragedia de 1999*. Caracas, Taurus-Santillana.

Vázquez, F. (2011) *Recuerdo de una tragedia: Vargas 1999–2009. Memoria, olvido y desastre en el estado Vargas, Venezuela*. BA Thesis. Caracas, Escuela de Antropología, Universidad Central de Venezuela.

Wilches-Chaux, G. (1998) *Auge, caída y levantada de Felipe Pinillo, mecánico y soldador o Yo voy a correr el riesgo*. Lima, LA RED-ITDG.

Wilches-Chaux, G. (1993) La vulnerabilidad global. In: Maskrey, A. (comp.). *Los desastres no son naturales*. Bogota, LA RED-Tercer Mundo Editores, pp. 11–44.

Wolf, E. (1999) *Envisioning Power: Ideologies of Dominance and Crisis*. Oakland, University of California Press.

Wolf, E. (1982) *Europe and the People Without History*. Berkeley, University of California Press.

Index

Note: Page numbers in *italics* indicate maps.

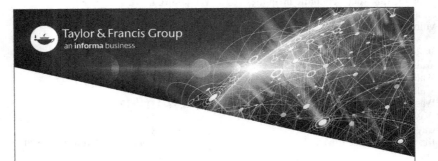

Printed in the United States
by Baker & Taylor Publisher Services